LIFE

Life is a compelling addition to the Darwin College Lecture Series, in which eight distinguished authors each present an essay from their area of expertise devoted to the theme of 'life'. The book forges connections among art, science and the humanities in a vibrant and thought-provoking collection which exposes both conventional and unconventional views on the meaning of life, the enigmatic boundaries between the living and the dead, and what may or may not follow afterwards. This collection arises from the Darwin College Lecture Series of 2012 and includes contributions from eight scholars, all of whom are held in esteem not only for their research, but also for their ability to communicate their subject to popular audiences.

Contributors:
Michael Akam, Frances Ashcroft, Chris Bishop, Mark de Rond, Clive Gamble, Ron Laskey, Robert Macfarlane and Michael Scott.

WILLIAM BROWN CBE is Emeritus Master of Darwin College and Professor of Industrial Relations in the Faculty of Economics at the University of Cambridge.

ANDREW FABIAN OBE is Fellow of Darwin College, Royal Society Professor and Director of the Institute of Astronomy at the University of Cambridge.

THE DARWIN COLLEGE LECTURES

These essays are developed from the 2012 Darwin College Lecture Series. Now in their twenty-eighth year, these popular Cambridge talks take a single theme each year. Internationally distinguished scholars, skilled as popularizers, address the theme from the point of view of eight different arts and sciences disciplines.

Subjects covered in the series include

26 LIFE
eds. William Brown and Andrew Fabian
pb 9781107612556

25 BEAUTY
eds. Lauren Arrington, Zoe Leinhardt and Philip Dawid
pb 9781107693432

24 RISK
eds. Layla Skinns, Michael Scott and Tony Cox
pb 9780521171977

23 DARWIN
eds. William Brown and Andrew Fabian
pb 9780521131957

22 SERENDIPITY
eds. Mark de Rond and Iain Morley
pb 9780521181815

21 IDENTITY
eds. Giselle Walker and Elizabeth Leedham-Green
pb 9780521897266

20 SURVIVAL
ed. Emily Shuckburgh
pb 9780521718206

19 CONFLICT
eds. Martin Jones and Andrew Fabian
hb 9780521839600

18 EVIDENCE
eds. Andrew Bell, John Swenson-Wright and Karin Tybjerg
pb 9780521710190

17 DNA: CHANGING SCIENCE AND SOCIETY
ed. Torsten Krude
hb 9780521823784

16 POWER
eds. Alan Blackwell and David Mackay
hb 9780521823717

15 SPACE
eds. François Penz, Gregory Radick and Robert Howell
hb 9780521823760

14 TIME
ed. Katinka Ridderbos
hb 9780521782937

13 THE BODY
eds. Sean Sweeney and Ian Hodder
hb 9780521782920

12 STRUCTURE
eds. Wendy Pullan and Harshad Bhadeshia
hb 9780521782586

11 SOUND
eds. Patricia Kruth and Henry Stobart
pb 9780521033831

10 MEMORY
eds. Patricia Fara and Karalyn Patterson
pb 9780521032186

9 EVOLUTION
ed. Andrew Fabian
pb 9780521032179

8 THE CHANGING WORLD
eds. Patricia Fara, Peter Gathercole and Ronald Laskey
9780521283281

7 COLOUR: ART AND SCIENCE
eds. Trevor Lamb and Janine Bourriau
9780521499637

6 WHAT IS INTELLIGENCE?
ed. Jean Khalfa
pb 9780521566858

5 PREDICTING THE FUTURE
ed. Leo Howe and Alan Wain
pb 9780521619745

4 UNDERSTANDING CATASTROPHE
ed. Janine Bourriau
pb 9780521032193

3 WAYS OF COMMUNICATING
ed. D. H. Mellor
pb 9780521019040

2 THE FRAGILE ENVIRONMENT
eds. Laurie Friday and Ronald Laskey
pb 9780521422666

1 ORIGINS
ed. Andrew Fabian
pb 9780521018197

Life

Edited by *William Brown* and *Andrew Fabian*

Darwin College, Cambridge

CAMBRIDGE
UNIVERSITY PRESS

University Printing House, Cambridge CB2 8BS, United Kingdom

One Liberty Plaza, 20th Floor, New York, NY 10006, USA

477 Williamstown Road, Port Melbourne, VIC 3207, Australia

314–321, 3rd Floor, Plot 3, Splendor Forum, Jasola District Centre, New Delhi - 110025, India

79 Anson Road, #06-04/06, Singapore 079906

Cambridge University Press is part of the University of Cambridge.

It furthers the University's mission by disseminating knowledge in the pursuit of education, learning and research at the highest international levels of excellence.

www.cambridge.org
Information on this title: www.cambridge.org/9781107612556

First published 2014

A catalogue record for this publication is available from the British Library

Library of Congress Cataloging in Publication data
Life / edited by William Brown, Andrew Fabian.
pages cm – (Darwin college lectures ; 25)
Includes bibliographical references and index.
ISBN 978-1-107-61255-6 (paperback)
1. Life. I. Brown, William Arthur, 1945– II. Fabian, A. C., 1948–
BD431.L4163 2014
113′.8 – dc23 2014012948

ISBN 978-1-107-61255-6 Paperback

Contents

Figures

Tables

Notes on contributors

Michael Akam studied Zoology at Cambridge and Oxford before research at the Laboratory of Molecular Biology and at Stanford. Returning to Cambridge, he became Director of the University Museum of Zoology in 1997 and Head of the Department of Zoology in 2010. He is a Fellow of Darwin College. His early work was on the genes that control development. More recently he has explored the genetic basis for animal diversity using our expanding knowledge of developmental genetics.

Frances Ashcroft FRS studied at Cambridge. She is currently a Royal Society Research Professor in the Department of Physiology, Anatomy and Genetics at the University of Oxford, and a Fellow of Trinity College, Oxford. Her research is on the process through which blood glucose concentration stimulates the release of insulin, and what goes wrong with this process in diabetes. She has published two popular science books, *Life at the Extremes* (2000) and *The Spark of Life* (2012).

Chris Bishop studied Physics at Oxford and Edinburgh and is now is a Distinguished Scientist at Microsoft Research, and a Fellow of Darwin College, Cambridge. He is also Vice President of the Royal Institution, and Professor of Computer Science at the University of Edinburgh. His research interests include probabilistic approaches to machine learning, as well as their application in a broad range of scientific and technological domains. His books include *Neural Networks for Pattern Recognition* (1995) and *Pattern Recognition and Machine Learning* (2006). In 2008 he gave the Royal Institution Christmas Lectures.

Mark de Rond studied at Oxford and is Reader in Strategy and Organization at Cambridge University Judge Business School and a Fellow of Darwin College. He studies teams of high performers by living with them under similar conditions. His books, which include *Strategic Alliances as Social Facts*

(Cambridge University Press, 2003), *The Last Amateurs: To Hell and Back with the Cambridge Boat Race Crew* (2008) and *There Is an I in Team: What Sports Coaches and Elite Artists Really Know about High Performances* (2012), have won a number of awards.

Clive Gamble FBA studied at the University of Cambridge. He founded and directed the Centre for the Archaeology of Human Origins at the University of Southampton where he is now a professor. He has undertaken research into the evolution of human society concentrating in particular on the Palaeolithic. His recent books include *Origins and Revolutions: Human Identity in Earliest Prehistory* (Cambridge University Press, 2007), *Archaeology: The Basics* (2008) and the international-award-winning *The Palaeolithic Societies of Europe* (Cambridge University Press, 1999).

Ron Laskey FRS CBE studied at Oxford and has recently retired from the Charles Darwin Chair of Animal Embryology in the University of Cambridge and the Directorship of the MRC Cancer Cell Unit in the Hutchison/MRC Research Centre at Cambridge. He is a Fellow of Darwin College. His main research interest has been the control of cell proliferation and why it goes wrong in cancer. He has written and recorded *Selected Songs for Cynical Scientists.*

Robert Macfarlane studied English at Cambridge and Oxford and is a Fellow of Emmanuel College and Senior Lecturer in the Faculty of English at Cambridge. He is the author of *Mountains of the Mind: A History of a Fascination* (2003), *Original Copy: Plagiarism and Originality in Nineteenth-Century Literature* (2007), *The Wild Places* (2007), *The Old Ways* (2012) and *Holloway* (2013). His books have won numerous national and international awards.

Michael Scott studied Classics at Cambridge and is now an Assistant Professor at Warwick University. He was formerly the Moses and Mary Finley Fellow in Ancient History at Darwin College. His publications include *From Democrats to Kings* (2009), *Delphi and Olympia* (Cambridge University Press, 2010) and *Space and Society in the Greek and Roman World* (Cambridge University Press, 2012). Committed to communicating about the ancient world to as wide an audience as possible, he has written and presented a number of television documentaries.

Preface and acknowledgements

The boundary between life and non-life has been the guiding principle for this interdisciplinary exploration of the notion of life. The chapters that follow start with cells, bio-electrical mechanisms, evolutionary processes and artificial intelligence. Then, in the social world, they consider work on the boundary of death, the way we have envisaged life in the distant past, the metaphor of ruined life, and how first humanity imagined going beyond life.

Cells are the minuscule bricks of life. Ron Laskey describes how living things are kept alive and healthy by the balancing of life and death among the trillions of cells of which they are made. Different functions require cells to have very different life expectations, from a few days to the whole life of the body. Each cell's birth and death is wholly altruistic. It is determined by what is needed for the best functioning of the body of which they are so tiny a part. The scale and complexity of what is required to keep a whole organism alive and healthy stretches our imagination. At the heart of every cell's birth is the process of division and thus replication of its DNA, an act of, in terms of man-made things, incomprehensible precision.

It is their role as electrical machines in the process of life that is the crucial aspect of cells for Frances Ashcroft. Electricity is the literal spark of life which informs and powers the muscles that allow an organism to function. Her account of physiologists' growing understanding of this mechanism focuses on the proteins, known as ion channels, which permit electrical charges to be released or inhibited in the interaction of cells and nerve circuits. These are fundamental not only to how life proceeds, but also to how a vast variety of toxins are created in nature both to attack and defend it. A deepening understanding of the many roles of ion channels is proving to be of great importance to medical science.

Moving from the bricks and control systems of life to whole organisms, the extraordinary achievement whereby organisms are evolved becomes apparent. Michael Akam discusses what the analysis of genomes is telling us about this process. It reveals how misleading the appearance of a plant or animal might be as an indicator of its forebears. The power of natural selection can create superficially similar life-forms from ones with profoundly different histories. The evolution of life has not been one towards ever-increasing complexity. The underlying genetic toolkit has not changed much over the whole period in which all living organisms have been evolving from a common ancestor. Co-operative interaction is key to this toolkit. Some relatively simple types of cell appear to have evolved from more complex cells, finding benefit in the division of their specific labour. There appears to be a logic embodied in the interaction of genes themselves which controls the evolution of life.

Despite the extreme complexity of life, the potential for simulating or even creating aspects of it has been an increasingly attainable ambition with rapid advances in computer science. Chris Bishop explores the scope for artificial intelligence around three seminal ideas of Alan Turing. He starts with the insights that mathematics has provided of the capacity of interactive systems to explain the development of structure in living systems. He then sketches the quest to create artificial intelligence, with its progression from expert learning to machine learning, and on to the construction of neural networks, exploiting the immensity of contemporary computing power. He ends with the emerging field of synthetic biology utilizing understanding of genetic codes to manipulate the fact that life is 'a system which manages information'.

The extraordinary experience of six weeks in a busy military field hospital in Afghanistan provided Mark de Rond with the opportunity to see how people work at a raw interface between life and death. He takes us into the worlds of the soldiers coping with killing, of the medics struggling to avert death, and of the photographers seeking to catch the images that will hold the world's attention. Success in these roles depends upon managing, as an individual, shocking contradictions, and upon seeing them as such, and not to be glossed over. Paradoxically, he suggests that this may best be done if team-working focuses less on maintaining inter-personal harmony and more on protecting a sharp awareness of the cruel contradictions inherent in the work.

In answering the question of what life was like in the ancient world, Michael Scott tells a vivid story of the uses and abuses of history. It is a story of how successive generations have chosen to portray a place and past where gems of insight and knowledge have gradually emerged from a mud of ignorance. Interpretations of life in ancient Greece have always been responsive to the attitudes and needs of subsequent worlds. He shows how the classical world has, over the subsequent centuries, been described, re-described, contested and idealized to fit whatever the current debates might have been. It was all so very long ago, and yet the remaining written and artistic evidence is so exceptionally rich. As a result, the process of debate over the nature of the ancient Greek civilization continues to be a powerful part of the intellectual life of our modern world.

A potent artistic device for the understanding of human life has been the imaginary exploration of its ruination. The fragile order of life and its structures degrades into weeds and jungle. An account of the gentle demise of an unpeopled city of Cambridge opens Robert Macfarlane's discussion of the way this literary tradition has evolved. Early 'ruinists', who celebrated the imagined triumph of nature over the works of man, gave way to a more melancholic or romantic approach. More recently, exemplified by the poet and naturalist Edward Thomas, the use of the metaphor has shifted to be neither hateful nor nostalgic. The challenge for our age is an accommodation with nature, rather than wishing either its eradication or its triumph.

The after-life discussed by the archaeologist Clive Gamble was not the physical one of graves and tombs. It was the more profound one of when the human mind started to go beyond the here and now. When did our forebears begin to imagine absent others, whether alive or dead? Such 'going beyond' involves not just living people but the material things in which we are enmeshed and which shape us as social beings. It implies that another person has a different perspective than you do; that they have a mind. He argues that the notion of an after-life emerged as a feature of mobile people, who lived at low population densities, fishing, gathering and hunting, but equipped with the means to think ahead. They did this materially through the metaphors of containers and instruments, and linguistically through signs and symbols. They created their own after-lives, as do we.

This collection of essays would not be complete without an acknowledgement to the many Members of Darwin College who facilitated the lecture series, and, in particular, to Richard and Ann King for their generous financial support, and to Janet Gibson, who brought order both to the contributors and to their manuscripts.

William Brown and Andrew Fabian

1 Life and death of a cell

RON LASKEY*

This article is not about prison reform, death in police custody or design of medieval monasteries. Instead the cells that it concerns are the living cells that make up our bodies. Most readers will be aware that estimates of the number of human bodies on the planet reached 7 billion in 2011 and none of us has difficulty recognizing 7 billion as a simply enormous number. Therefore it may come as a surprise to discover that 7 billion cells would make up only the terminal joint of my index finger (Figure 1.1). The total number of cells in the human body is best estimated at 100 trillion, 10^{14}. The inevitable conclusion from this is that cells are extremely small, with occasional conspicuous exceptions, like an ostrich egg, which begins as a single fertilized egg and is thus an enormous single cell.

This chapter starts by a simple introduction to the beauty and fascination of living cells. They are responsible for building all the tissues of the body, including blood, nerves, muscle, bone, yet they are all formed by progressive specialization from the cells generated by division of a single fertilized egg. This poses two extraordinary challenges. The first is the nature of the molecular mechanisms that allow cells to diverge and to specialize to fulfil particular functional niches, but the full details of these mechanisms lie outside the scope of this chapter. The second challenge is that of producing stable and balanced numbers of each type of cell within the body. How are the ratios of blood cells to nerve cells or cells that line our gut balanced and managed? This question is made all the

* I am grateful to Sally Hames for help with preparing this chapter. I thank Matthew Daniels, Jackie Marr and Peter Laskey for providing figures.

This lecture is dedicated to César Milstein, 1927–2002, inventor of monoclonal antibodies and Nobel Laureate in Physiology or Medicine 1984, and who recuited me to Darwin College.

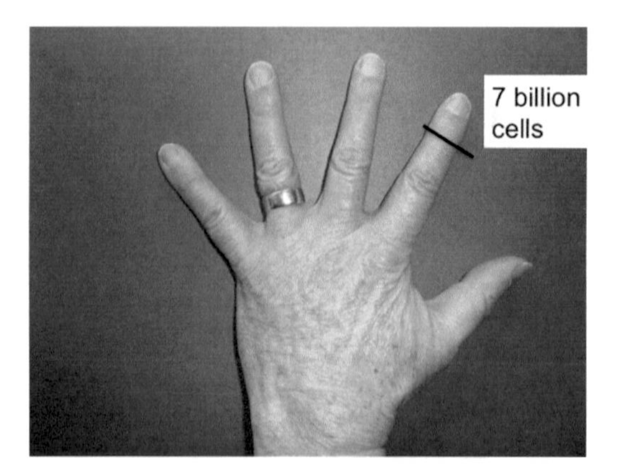

FIGURE 1.1 The number of humans on the planet is 7 billion, yet 7 billion is approximately the number of cells in just the terminal joint of one finger.

more acute by the fact that different types of cell persist in the body for very different lengths of time. Thus the cells that line our intestines, or the cells that line the ducts that carry digestive juices from our pancreas into the gut, survive for only a few days before they are replaced by new counterparts. In contrast most of our nerve cells persist throughout our adult lifetime. Although some types of nerve cell can be formed during our lifetime, others cannot and most remain with us throughout adult life. This poses an extraordinary challenge of bookkeeping and management of cell production and replacement.

After an elementary introduction to the workings of the living cell, this chapter will consider how new cells are formed by the process of cell division, an area in which our understanding has exploded within the last thirty years. It will then argue that death can be good for you, when it is cell death. Programmed cell death is a phenomenon that plays a crucial role in maintenance of the body structure and organization. Not only are worn-out cells constantly replaced and renewed by cell division, but programmed death of specific cells is a mechanism that sculpts the human body and plays an important part in determining the form of our bodies. However, with both cell division and cell death influencing the numbers of each cell population, co-ordination and regulation of cell population sizes becomes crucial. The nature of this challenge will be described and the ways in which it is managed will be outlined.

The two final sections will consider what happens when this tightly managed balance breaks down. Excessive cell proliferation, or inadequate cell death, can both contribute to cancer. These imbalances can be triggered in many different ways, resulting in uncontrolled growth of a selfish cell population. Conversely excess cell death without renewal or replacement can result in a range of degenerative diseases. Some of the most prevalent of these are the neurodegenerative diseases including Alzheimer's and Parkinson's diseases. Both of these, and several other neurodegenerative diseases, arise from disorders of protein folding. Proteins should fold into precise three-dimensional structures but then be unfolded and degraded when they are damaged or have served their purpose. However, errors in protein destruction can give rise to fragments that are capable of forming insoluble aggregates with seriously damaging consequences for the cells that contain them or are surrounded by them. The build-up of toxic insoluble protein aggregates in or around nerve cells can be responsible for death of those nerve cells and several important neurodegenerative diseases.

The living cell

Figure 1.2 illustrates the extraordinary beauty of cells. It is a photograph of a single cell viewed in a fluorescence microscope after three different cellular components have been stained red, green or blue. The blue stain reveals DNA packaged within the cell nucleus, which serves as the information archive of the cell. The red stain reveals mitochondria, the powerhouses of the cell that metabolize ingested food and convert it into a form of energy that can be used throughout the cell. In addition, mitochondria play an essential regulatory role in the control of cell death. The green stain reveals components of the cytoskeleton, namely microtubules that are essential for distribution of other large components around the cell. They play a central role in the process of cell division and the segregation of chromosomes to the two progeny cells that arise from cell division.

The packing of DNA within the cell nucleus is truly remarkable. Each cell nucleus contains 2 m of DNA and this encodes the information for the structure and function of all the cells in the body. Yet these 2 m of DNA

Table 1.1. *The problem of packing DNA into the cell nucleus. The extraordinary scale of the problem is seen more clearly by a scale model in which all dimensions are increased 1-million-fold.*

Dimensions of DNA and the cell nucleus, actual and magnified 1 million times		
		$\times 10^6$
DNA diameter	2 nm	2 mm
DNA length in nucleus	2 m	2000 km
Diameter of nucleus	5 μ	5 m

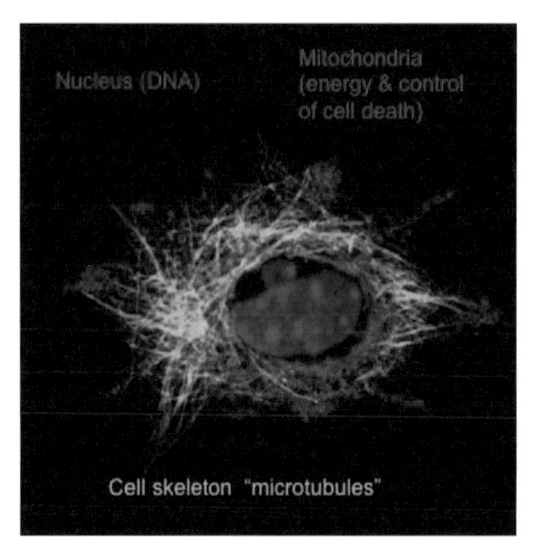

FIGURE 1.2 A fluorescent micrograph of a human cell stained to reveal DNA in the cell nucleus (blue), mitochondria that are the power units of the cell (red) and microtubules that are dynamic components of the cell skeleton (green). Image reproduced with permission from Peter Laskey.

are packed into a cell nucleus of only a few microns in diameter in such a way that all the DNA is available for duplication to produce the nuclei of the progeny cells after cell division. The scale of this packing problem can be understood more clearly by a scale model in which everything is increased 1-million-fold, as shown in Table 1.1. On this scale the nucleus would have a diameter of about 5 m and DNA would have a diameter of 2 mm, the diameter of thin string, but the length of DNA enclosed in each of these 5-metre nuclei would reach from St Pancras to St Petersburg, namely 2000 km. Remarkably this must all be packed in in such a way that it is accessible for copying in readiness for cell division. A second example illustrates the extraordinary length of DNA packaged in each cell. I stated that the terminal joint of my index finger contains approximately 7 billion cells. All five complete digits from one hand would have approximately 100 billion cells and each of these contains 2 m of DNA. The conclusion is, therefore, that if DNA was extracted from all five digits of one hand and joined end to end the resulting DNA molecule would be 200 million km long, enough to reach the sun. Remember that this example is not a scale model; it is reality.

The capacity for information storage within DNA is remarkable and is superbly illustrated by an exhibit in the Wellcome Collection at the Headquarters of The Wellcome Trust in London. The four bases A, C, T and G, read in groups of three, encode information in DNA. The exhibition in the Wellcome Collection prints out the human genome in a normal type font and it is bound into a large volume on display with the letters A, C, T and G repeated many times on each page. The volume that is opened on display is the size of a large encyclopaedia but remarkably it is one of 118 such volumes that represent the complete human genome. They occupy a large bookcase stretching from floor to ceiling, yet this is the information encoded in each copy of the human genome and stored in each of the 100 trillion cells in our bodies. In micro-fabrication, we clearly still have a lot to learn.

Figure 1.3 shows how information flows from DNA in the cell nucleus to produce the many types of specific protein that are made outside the nucleus, in the cytoplasm. The information in one strand of DNA is copied into a related molecule, called RNA, that is exported from the nucleus through pore complexes in the nuclear envelope, to the cytoplasm. There it serves as a template to be translated by specialized machinery, called ribosomes, to produce the many proteins of the cell. It is the proteins that then build the structure and do the work, including catalysing the many biochemical reactions within the cell. In summary, the cell is an extraordinarily organized information storage and interpretation machine that

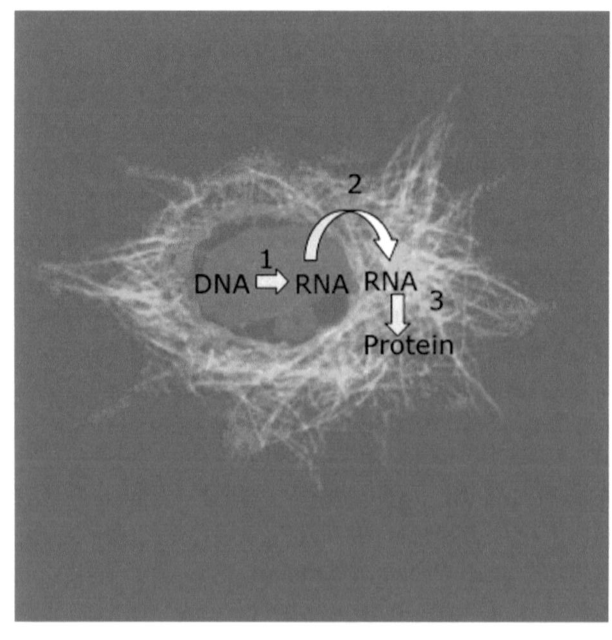

FIGURE 1.3 Information flow from DNA in the cell nucleus for synthesis of proteins in the cytoplasm.

encodes vast amounts of information in very small spaces and yet selectively retrieves subsets of that information for expression in different types of cell. The reason why cells differ from each other in the body is not that they contain different DNA, but that they copy different subsets of the DNA into RNA and therefore make different proteins. A sophisticated network of controls determines which regions of DNA, which genes, are copied into RNA to make proteins. After a subset of genes have been copied from DNA into RNA, there are further levels of control before the proteins are made. However, these necessarily lie outside the scope of this chapter.

Cell division

In 1855 a Prussian pathologist, Rudolph Virchow, wrote: 'Omnis cellula e cellula' [All cells come from cells], indicating that all cells are

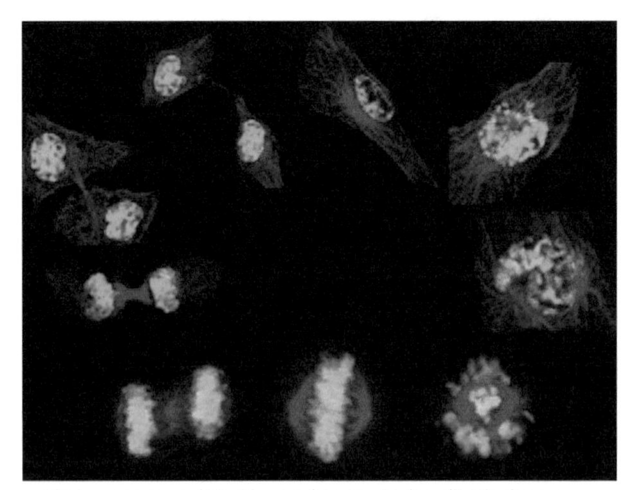

FIGURE 1.4 Stages of cell division starting with a single cell at 1 o'clock and proceeding clockwise to form 2 cells at 11 o'clock. Chromosomes are green and microtubules are red. Image reproduced with permission from Matthew Daniels.

derived by division of existing cells rather than assembled de novo. Cell division is the most dramatic event that takes place within a cell's lifecycle. It involves condensation of the long DNA threads in the nucleus into discrete, microscopically visible, chromosomes and their division longitudinally by a mechanical spindle apparatus. Between consecutive divisions DNA is dispersed throughout the cell nucleus, but complexed to specific proteins that neutralize the acidic charges of its many phosphate groups and that enable it to be folded into a hierarchy of structures in the nucleus. This is the state in which DNA is copied to make RNA and also duplicated to make more DNA. Once the DNA has been copied to produce two complete copies, it then becomes coiled into visible chromosomes for division between the two progeny cells. In human chromosomes the DNA is compacted 10,000-fold allowing it to be divided in two within the boundaries of the cell.

The mechanism of cell division requires attachment of the chromosomes to the ends of microtubules (stained green in Figure 1.2 and red in Figure 1.4), with one side of each chromosome attached to microtubules extending from each pole of the spindle. Microtubules have functions in cells between divisions. They serve as dynamic rails along which other cellular components are distributed, but they become fundamentally reorganized during division, by two microtubule organizing centres, from which they grow. The two halves of each chromosome become attached to the ends of microtubules ready to be divided longitudinally for distribution into the two progeny cells. At this time each of the two DNA molecules arising from DNA synthesis is coiled into one of the two parallel halves ('chromatids') and these are held together by ring shaped proteins called cohesins that appear to physically hold the two halves together until all chromosomes are equally attached to both spindle poles. Division is delayed until each chromosome is firmly attached to both poles of the microtubule spindle. This delay causes chromosomes to align in a flat plate experiencing equal forces pulling them towards each pole, and only then can they split longitudinally for division, a process that is achieved by cleaving the cohesin protein rings that link the two new DNA molecules together. In addition, chromosome condensation and division of the chromosome in two are both delayed until all DNA synthesis is complete. Incomplete DNA synthesis would result in chromosome breakage when the chromosomes are pulled apart during cell division. The 'checkpoint' mechanisms that delay events during cell division to ensure that each event is complete before the next one starts are crucial in maintaining the integrity of the genetic information in each cell. Maintaining the integrity of the genome in these ways is equally crucial to prevent cancer. Mutations in the genes that regulate cell proliferation either positively or negatively are the raw materials for the pseudo-Darwinian selection that selects for the fastest dividing cells in cancer. I say 'pseudo-Darwinian' because, as explained later, real Darwinian selection acts at the level of the individual, rather than at the level of the selfish cancer cell. True Darwinian selection selects against excessive proliferation of selfish cancer cells and thus against genetic instability and other causes of cancer. Instead it selects for genome stability and checkpoint mechanisms that defend the individual against instability of the genome and thus defend against the risk of cancer.

Not only is it essential to ensure that all of the DNA has been synthesized to make two complete copies before the cell can divide, but it is also essential that no part of the DNA is copied twice. The genetic imbalances that arise from either incomplete DNA synthesis or repeated DNA synthesis of the same piece of DNA both provide the raw materials for carcinogenic changes resulting in cancer. Duplicating the immensely long threads of DNA that exist in our cells poses another serious problem. DNA is a double helix. One strand wraps around the other once for each ten letters of the genetic code. Therefore as a human cell contains 6×10^9 letters (3×10^9 from each parent), there are 6×10^8 (600 million) helical turns to be removed to unwind the DNA strands before they can be separated into the progeny cells. This generates important roles for two classes of proteins that will feature later in this chapter. Unwinding one strand from the other requires unwinding activities called DNA helicases, because they unwind the DNA double helix, but as they do so they simply move the helical twists from the DNA that is undergoing synthesis and pass them on to the DNA that has not yet been synthesized generating a serious topological tangle. Fortunately the cell has two classes of activity that ingeniously solve this problem. They have the ability to remove torsional stress from DNA by cutting and rejoining it. Unfortunately they have the off-putting names of 'DNA topoisomerases', but what they do is remarkable. Type 1 topoisomerases cut one strand of DNA, hold on to the end by a chemical bond and pass the other strand through a gap that they have made, removing helical twists one at a time. Each time they do this they repair the gap that they made, restoring the DNA to its original structure but with less torsional stress. Type 2 topoisomerases do something even more remarkable. They cut through both strands of the DNA, holding on to the cut ends, and pass a second double helical DNA molecule through the temporary gap that they have made. This can untangle the most troublesome tangle. Both these extraordinary activities will feature later in the chapter.

Death can be good for you, when it's cell death

During the time it will take to read this chapter, approximately a billion of your cells will have died. Fortunately approximately a billion will also have been produced by new cell division. The balance of these processes is remarkable as the population of each of the many types of cell in our bodies is very stably regulated. Cell death is not a random or haphazard event. It is precisely regulated and executed with extraordinary efficiency. Programmed cell death (otherwise known as apoptosis) is mediated by a cascade of molecular assassins that destroy the cell completely so that only residual remnants remain to be consumed by its neighbours. Once a signal for a cell to die is received, it is amplified by a series of enzymes that culminate in chopping the essential proteins of the cell into non-functional fragments as well as cleaving DNA into irreparable fragments. Mitochondria (stained red in Figure 1.2) modulate the damage level and amplify the death signal in a condemned cell. They do this by releasing the protein cytochrome C out of the mitochondrion. Its release activates formation of a disc-shaped protein complex called the 'apoptosome' or 'wheel of death' that accelerates a cascade of destructive enzymes, resulting in the systematic disintegration of the cell. Apoptotic cell death is so decisive that there is no risk of a genetically damaged cell persisting in a state that could allow it to divide and become cancerous. Damaged or wounded cells are destroyed absolutely and irreversibly, as seen in progress in Figure 1.5. In addition to contributing to turnover and renewal of cells in the mature body, programmed cell death also plays an important role in shaping and sculpting tissues. For example, the human hand is sculpted from a flat paddle in the embryo by death of cells between the digits rather than just by growth of the digits themselves. They are literally carved out of a flat paddle-shaped limb, just as a sculptor would carve them from a block of stone (Figure 1.6).

In addition to the tight regulation of a cell's lifespan by programmed cell death, the length of time and number of generations for which a cell is able to divide are also tightly regulated. When cells are grown in culture they normally divide for only a limited number of generations and then stop. They become senescent. The timing of senescence is sometimes correlated with erosion of the specialized ends of chromosomes called telomeres. These have specialized DNA structures that are not duplicated by the normal DNA synthesis machinery, but extended by a dedicated enzyme called telomerase. This enzyme becomes inactivated in mature cells as they differentiate or senesce. It is maintained in an active state in stem cells or reactivated in cancer cells.

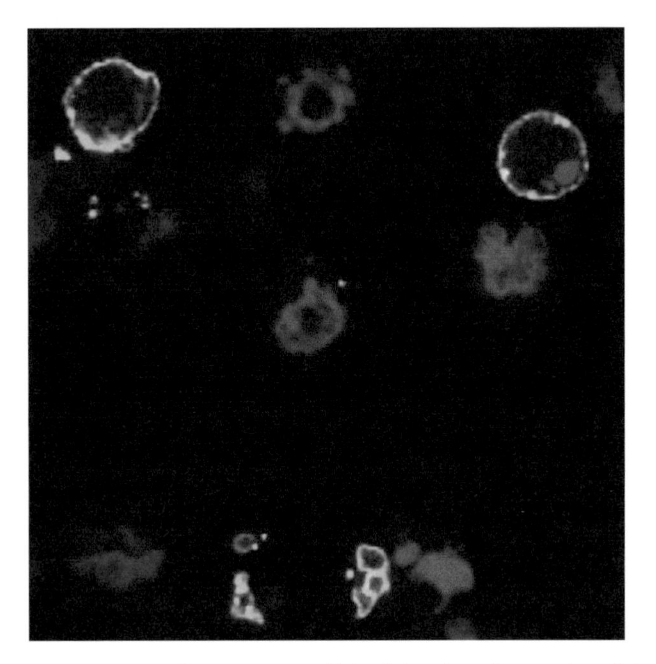

FIGURE 1.5 Programmed cell death involves destruction of the DNA stained red and disruption of the cell membrane stained green. Image reproduced with permission from Jackie Marr.

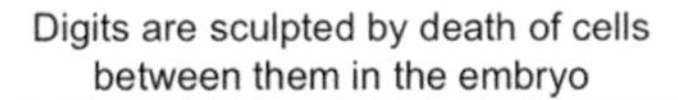

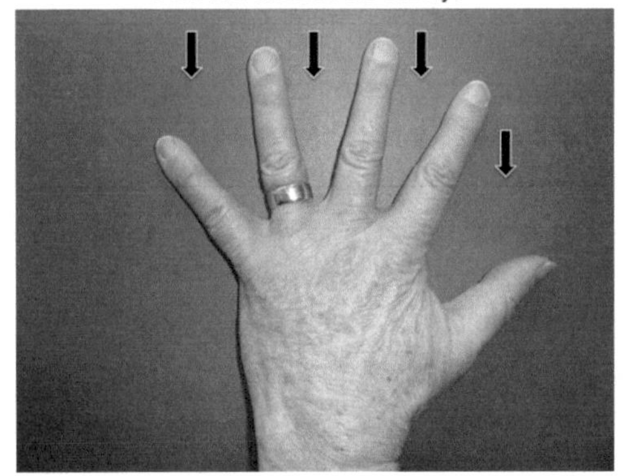

FIGURE 1.6 Programmed cell death sculpts the human hand in the embryo by selective death of cells between the digits.

Table 1.2. *Different types of cell in the human body are renewed at fundamentally different rates.*

Cell survival times in the human body	
Gut lining	Days
Pancreatic duct	Days
Skin cells	Weeks
Red blood cells	Months
White blood cells	Days OR months
Nerve cells	Lifetime (mostly)
Odour receptors	Months
Retinal receptors	Lifetime
'Hair cells' of cochlear	Lifetime

Cell renewal

Table 1.2 shows that different types of cell can persist in the body for radically different times. At one extreme, most of our nerve cells are not renewed but persist throughout our adult life. This is also true of the 'hair cells' that detect sound in the cochlea of the ear, or the photoreceptors that detect light in the retina of the eye. Replacement and renewal after damage are not options. In contrast, odour receptors are renewed every few months, as are red blood cells, whereas skin cells are replaced every few weeks. At the opposite extreme from nerve cells are the cells that line the gut, or the ducts that carry digestive juices from the pancreas to the gut. Each of these is replaced in a matter of a few days, reflecting their exposure to environmental damage.

As can be seen from these examples, patterns of cell renewal are very tightly regulated. Cells divide only in the right place at the right time, with an important exception. The exception of course is cancer, which is characterized by uncontrolled cell proliferation. In most of our tissues, most of the cells are programmed to undergo cell death after a predetermined time. The exception to this rule defines stem cells. Stem cells are immortal and persist throughout the lifetime of the individual. They divide relatively rarely to produce progenitor cells that then maintain the cell population for a limited length of time producing very many

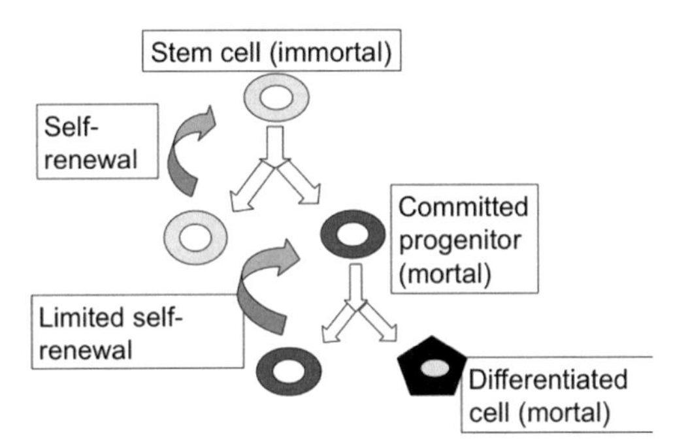

FIGURE 1.7 Only stem cells are immortal. When they divide they renew themselves and produce long-lived, but mortal progenitor cells. These, in turn, renew themselves for a limited period and produce specialized 'differentiated' cells.

differentiated cell products as well as renewing themselves by self-renewal for a limited but sometimes extensive time. These patterns are illustrated in Figure 1.7. The rate of production of new cells can be modulated, for example by high altitude in the case of red blood cells. High altitude induces a hormone, erythropoietin, which serves as a growth factor to induce production of more red blood cells, thus increasing the oxygen carrying capacity of the blood. Of course there are other ways of achieving this in addition to high altitude, notably direct injection of erythropoietin (EPO), an issue that is topical in this Olympic year.

Networks of growth factors, secreted by neighbouring cells, maintain the balance between cell types within the body. The effects of growth factors can also be modulated by varying the number of their receptors on the cell surface, so that a fixed concentration of growth factor can produce different levels of response because of variations in the sensitivity of the recipient cell. Growth factors not only determine cell number, but they can also determine cell type during the development of the embryo, as cells respond to their position in a gradient of growth factors by differentiating to specialize in specific ways. Stem cells divide relatively rarely, which minimizes their risk of genetic changes during DNA duplication. One way in which this is achieved is by the designation of stem cell niches. These are stem cell pockets in which stem cells can persist and maintain their stem cell identity through physical contact with neighbouring cells that transfer growth factors to them directly by intercellular contact.

This contributes to the maintenance and form of the tissue because then stem cells can only proliferate within a precise structural framework. The restriction of immortality to a small minority of cells in the body is an important defence against the risk of cancer as will be explained in the next section.

Cancer

One of the principal steps in the development of cancer is the breakdown of the regulations that control cell renewal. Cancer cells ignore or bypass the strict social conventions that maintain cell number, resulting in the unrestrained growth of a tumour. Two classes of gene should normally regulate cell growth. The first are oncogenes, which control cell proliferation by switching it on. They are often overactive in tumour cells. The second class of genes that control cell proliferation are the tumour suppressor genes. In general, they either control cell proliferation by switching it off, or they maintain genetic integrity by repairing DNA damage before it becomes fixed in DNA by new DNA synthesis. They are often inactivated in cancer cells. Examples include the breast cancer susceptibility genes BRCA1 and BRCA2 or the retinoblastoma gene Rb, whose mutation or loss results in an inherited form of blindness, retinoblastoma. Disrupted control of cell proliferation is not the only contributor to tumour formation. Other steps include invasion to colonize new sites, formation of new blood vessels to provide a blood supply to the growing tumour and, crucially, immortalization so that tumour cells evade cell death. Clearly if tumour cells were programmed to die at a fixed time, like most of the cells in the body, tumours would shrink spontaneously or never arise at all. To become a tumour, cells must effectively become immortal. This can arise in either of two ways. Either it can arise by changes to a stem cell that is already immortal so that the stem cell divides too frequently, or it can arise by genetic changes that make a tumour cell become immortal, resembling stem cells. Ironically, although many viruses have evolved to kill cells, some viruses such as papilloma viruses can immortalize cells by inactivating the cell death machinery or its controls. This negates the cell's attempt to terminate the viral infection by its own altruistic death. Instead, it keeps the infected cell alive so that it continues to produce infectious viruses for much longer. A remarkable account of the best-known example of this can be read in Rebecca Skloot's account of *The Immortal Life of Henrietta Lacks.* HeLa cells from the cervical cancer of Henrietta Lacks have been maintained in laboratories for more than sixty years since Henrietta Lacks died.

If cancer cells are immortal this appears to raise an important paradox. Why are immortal cancer cells more vulnerable to cancer drugs than their mortal neighbouring normal cells? The answer lies in checkpoints that ensure the damaged cells stop making DNA or dividing until they have been repaired. These mechanisms ensure the genetic integrity of cells by waiting while any genetic damage is repaired to prevent it from being passed on to dividing cells. Furthermore, if the damage is too extensive to be repaired effectively, then cell death is triggered and the cell is rapidly dispatched. However, to become a cancer cell it is necessary to switch off regulatory mechanisms that control cell proliferation including the checkpoints that delay DNA synthesis or division until the damage has been repaired. Therefore cancer cells persist with proliferation in the face of damage, whereas normal cells pause to repair that damage. Therefore when damage is inflicted on the cell by chemotherapeutic drugs, the normal cell will pause its proliferation until that damage is repaired, whereas the tumour cell will continue proliferating and sustaining so much damage that it becomes genetically unstable and dies. Unfortunately the increased genetic instability of cancer cells is also their escape route from chemotherapy. If all cancer cells were the same, they would respond similarly to chemotherapy and tumours could be cured. But because cancer cells are genetically less stable than their counterparts, they continue to undergo change, generating the raw material for selection of surviving resistant cells that re-establish the tumour anew. The selective 'advantage' of the fastest dividing cells that allows them to outgrow their neighbours and the selection for the most drug-resistant cells during chemotherapy resemble Darwinian selection of the fittest organisms. However, there is a fundamental difference because perversely, selection for the fastest-growing cells, i.e. the most rapidly growing tumour, selects against survival of the individual. As true Darwinian selection acts at the level of the breeding individual, it selects against selfish excessive cell proliferation. Thus it selects against cancer, helping to explain why so few cells among the 100 trillion in our bodies divide selfishly resulting in cancer.

As explained above, the double helical structure of DNA requires unwinding of one DNA strand from around the other in order for it to be copied into two new DNA helices. The local unwinding that this causes generates twist and tangling in the other parts of the DNA molecule and replication can only continue if those twists are removed. The two remarkable activities that address this problem were introduced earlier as Topo 1 and Topo 2 (introduced earlier as DNA topoisomerase 1 and DNA topoisomerase 2, which for convenience can be called Topo 1 and Topo 2). Both of them are important targets for chemotherapeutic drugs. For example Irinotecan is an inhibitor of Topo 1. As explained earlier, Topo 1 relieves torsional strain in DNA by nicking one of the two strands, holding tightly to the cut ends and passing the other strand through the nick. Irinotecan allows Topo 1 to nick the DNA but does not allow it to rejoin the nicked strands. Therefore it causes stable single strand breaks in DNA and prevents further DNA replication. Doxorubicin, Epirubicin and Daunorubicin act on Topo 2 which, as explained earlier, cuts through both strands of a DNA double helix, holds tightly to the cut ends and passes a second entire double helix through the gap. In this way it plays a crucial role in allowing the two progeny helices formed by duplication to separate so that they can be pulled apart during the process of cell division. The three Topo 2 inhibiting drugs mentioned above inhibit Topo 2 after it has cut through the DNA, but before it has rejoined the severed ends, resulting in breaks across both strands of the double helix. Once again this causes an inability of the cell to replicate and divide, resulting in preferential death of the tumour cells.

Just as activities that relieve torsional strain in a DNA molecule can be used as targets for cancer chemotherapy, so the activity that unwinds one strand from the other to allow them to be duplicated can also be used as a diagnostic tool to detect cancer earlier. One of the most immediate things that can be done to improve the survival of cancer patients is to detect cancer earlier in the course of the disease. Then existing treatments will become more effective. The particular diagnostic challenge we have attempted to address using this approach is the cervical smear test. The Papanicolaou smear test is an enormously important public health measure, because it can detect pre-cancerous cells before they have spread to remote sites, so before they have become too difficult to treat. However, as the normal cells are stained the same colour as the abnormals and differ only in subtle differences of shape, the Papanicolaou test is notoriously difficult to read and errors in reading it are a popular topic for adverse press coverage. We have attempted to use antibodies against proteins that unwind the two strands of DNA during DNA synthesis to stain proliferating cells a different colour from non-proliferating cells. The proteins that do this in plant and animal cells are called MCM proteins (an abbreviation for minichromosome maintenance proteins). They are abundant in the nuclei of proliferating cells but absent from non-proliferating or terminally differentiated cells. We have tested the ability of antibodies against MCM proteins to highlight abnormal cells in cervical smears by staining them a different colour from the normal cells in the smear, making them much easier to see. This technology has been subjected to extensive clinical trials at multiple centres in the United States, conducted by Becton Dickinson. It has resulted in one commercial product that highlights abnormal cells in clinical specimens, produced by Becton Dickinson and called ProExC. This preparation is not designed to be used as a frontline screening tool yet, but only to reinforce existing screening procedures. However, a second-generation reagent is in trials at the moment to test its suitability for use as a frontline cervical-screening method. The same approach has shown promising preliminary results for detecting cells recovered from other sites at which tumours form, such as bladder, colon and lung.

The two examples of applying knowledge of cancer-cell biology for patient benefit that are given above are just two of many advances that have arisen from studying the cell biology of cancer. There is now a steadily increasing emphasis on exploiting the knowledge of the life and death of cancer cells to deliver benefits for the patient.

Neurodegeneration

Just as excessive cell proliferation or insufficient cell death can contribute to cancer, so excessive cell death accompanied by insufficient cell renewal can contribute to a range of degenerative human diseases. Foremost

Table 1.3. *Approximate numbers of patients with different types of neurodegenerative diseases in the UK. Estimates for Parkinson's disease and Huntington's disease vary; minimal estimates for these are shown.*

Estimated numbers of neurodegenerative disease patients in the UK	
Alzheimer's disease	450,000
Parkinson's disease	>127,000
Huntington's disease	>6700
CJD	approx. 100 new p.a. (probable and definite cases)
vCJD	approx. 1 new p.a. (176 deaths since 1990)

amongst these are the neurodegenerative diseases including Alzheimer's disease and Parkinson's disease. Both of these result from excessive death of specific classes of nerve cells. In both cases the cause shares similarities with certain other neurodegenerative diseases including Huntington's disease and prion diseases such as CJD (Creutzfeld-Jakob disease). Table 1.3 summarizes the frequency of these conditions in the UK.

In all of these cases excessive nerve cell death is caused by formation of abnormal protein aggregates. Normally proteins should fold into precise three-dimensional structures, which in many cases allow the protein to be soluble or to form correctly ordered structures such as fibres. However, in the cases of the neurodegenerative diseases listed here, fragments of specific proteins aggregate together to form insoluble structures either within the cell, as seen for Parkinson's disease and Alzheimer's disease, or outside the cell as seen for Alzheimer's disease. The mis-folded protein in Parkinson's disease is called α synuclein. In the case of Alzheimer's disease, two kinds of protein aggregates are seen. The first, called neurofibrillary tangles, are intracellular aggregates of a protein called tau that modulates the formation and stability of microtubules (microtubules are stained green in Figure 1.2 and red in Figure 1.4) and that therefore is important for the maintenance and control of transport of other cargoes along microtubules through the long 'axons' that extend from nerve cells. The second type of aggregates are called amyloid plaques and they are formed outside the cell from a protein fragment called β amyloid that is derived from a protein that normally resides in the cell membrane. β amyloid fragments fold in a fundamentally different way from their precursor protein, forming insoluble aggregates between the nerve cells that produced them. The relative importance of intracellular tau protein aggregates and extracellular amyloid plaques in Alzheimer's disease is still hotly debated, though it is clear that both structures form and are associated with the disease.

The identity of the nerve cells that die is characteristic of the specific neurodegenerative disease. Thus in Alzheimer's disease nerve cell death affects the cerebral cortex that is involved in memory. In contrast, in Parkinson's disease the affected nerve cells are found in parts of the midbrain and they are cells that control movement. Specifically they are cells that produce the neurotransmitter dopamine that is involved in regulating movement, hence the treatment of Parkinson's disease patients with an orally administered dopamine precursor L–DOPA or with dopamine agonists.

An area of intense current investigation is the role of protein degradation in formation of the insoluble aggregates found in neurodegenerative diseases. Damaged or worn-out proteins are routinely replaced by mechanisms that mark them and then destroy them, in much the same way that dead cells are destroyed and replaced. Protein turnover is an important aspect of cell function, even in those cells which themselves persist throughout the lifetime of the individual. For example, although photoreceptors in the eye may last for full adult life, the proteins they contain can be turned over systematically and replaced by newly synthesized equivalents. Degradation of proteins is a highly regulated process that is carried out by specialized degradation machinery within the cell. The proteins that aggregate to form the toxic aggregates in neurodegenerative diseases appear to be misfolded or incorrectly degraded. Thus unfolded proteins can stick together rather than forming soluble independent entities. Similarly, the products of partial degradation of proteins can also aggregate unless they too are cleared and removed quickly by the protein degradation machinery. Errors in the organized degradation and replacement of old proteins are implicated in the formation of the toxic aggregates that lead to neurodegenerative diseases and efforts are currently underway to identify drugs that will prevent the formation, or cause the dissociation, of the toxic aggregates that cause so much trouble in the form of neurodegenerative diseases.

The tragic consequences of neurodegenerative diseases have rightly focused attention on the potential of stem cell research to provide future treatments. Nerve cells arise by division and differentiation of neural stem cells. Hence it should be possible in principle to reintroduce neural stem cells to the sites of neurodegeneration in the hope that they will then proliferate and differentiate to replace the neurons that have been killed. In the case of Parkinson's disease the objective appears relatively straightforward conceptually, though of course a fear remains that the capacity of stem cells to proliferate could generate uncontrolled growth and hence tumours. Awareness of that potential adverse outcome is of course a major consideration in attempts to exploit stem cells for such purposes. However, while it is clear that reintroduction of neural stem cells into patients requires appropriate caution, nevertheless it also provides hope for potential therapies for a range of neurodegenerative diseases and so it remains an area of active and potentially exciting research.

Conclusions and prospects

The social organization of the cells in our bodies is quite remarkable. Selfish, antisocial behaviour by one cell dividing repeatedly without restraint will lead to cancer, so it is remarkable that these events occur so rarely. If antisocial behaviour in the human population was as rare as antisocial selfish behaviour within the populations of cells that make up our bodies, then prison overcrowding would be a thing of the past as there would be far less than one prisoner on the planet at any one time. It is extraordinary how well disciplined and organized cell populations are. Millions of years of natural selection have allowed our bodies to develop principles of organization that minimize the risk of cancer, but also minimize the risk of excessive cell death without replacement.

We now understand in ever-increasing detail how cell populations are regulated, how the production of growth factors by one type of cell in one type of location can determine the rates of proliferation of another type of cell in a different location. We understand how the responsive cells react to the presence of a stimulatory growth factor and how those messages are transmitted from receptors at the surface of the cell deep into the interior of the cell to trigger proliferation. These are events that we can describe in ever-increasingly accurate molecular detail, but that knowledge does little to decrease our admiration for the extraordinary sophistication of the regulatory mechanisms involved. The balance between cell division and cell death within a specific cell population remains remarkable. In some cases it is relatively easy to understand, such as wounding the skin, which releases growth factors that stimulate growth of new skin cells to heal the wound. Similarly increased demand for oxygen such as that experienced at high altitude stimulates the body to produce the soluble hormone erythropoietin that induces the production of further red blood cells. Alternatively, damage to the liver whether mechanical or self-inflicted through alcohol abuse, can induce division of liver cells to replace the cells killed by the damage. Unfortunately the capacity for liver regeneration is limited and can be exceeded by persistent alcohol excess. Nevertheless each of these examples is a relatively simple concept to understand. It becomes more difficult to understand how the production of differentiated cells by division of stem cells and their progenitor offspring can lead to tissue homeostasis, precise balance between cell production and cell destruction in a stable cell population that neither expands, nor contracts. Yet this is happening in each of our tissues throughout our bodies throughout our lifetimes. The structural organization of tissues goes some way towards explaining this. For example, the proliferating zone in the outer layer of skin, the epidermis, is restricted to the tiers of cells that are adjacent to the lower layer, the dermis, so that contact with the dermis is required for cell proliferation. This provides a very simple method of ensuring that the right amount of epidermis is produced, not too many cells and not too few. Similarly, in the crypts of the intestine the geometry of those test-tube-like crypts specifies where cells may or may not proliferate. They proliferate around the wall of the deepest third of the cylindrical crypt and as they move up the crypt they cease proliferation. The need for physical contact with the substrate and cessation of cell division when cells become overcrowded on that surface substrate provide an effective and relatively simple method of regulating cell populations.

However, problems arise when cell death occurs within cell populations that are not naturally renewed such as nerve cells, resulting in neurodegenerative disease. Conversely problems arise when cells undergo changes that enable them to proliferate when they are not in contact with a supporting surface and therefore they are able to increase their number irrespective of the surface area of their substrate. This provides a first step towards the onset of cancer. I have written in more detail elsewhere about the molecular changes that contribute to cancer (Laskey 2004) but for the purpose of this discussion two key changes are the ability to divide without signals from growth factors or contact with a supporting surface and the ability to become immortal. As explained above only stem cells should be immortal, but genetic changes in a stem cell inducing it to keep dividing more frequently, or genetic changes in a mortal cell that induce it to become immortal like a stem cell, result in cancer.

In contrast to cancer, which arises from excess cell proliferation and insufficient cell death, neurodegenerative diseases like Alzheimer's disease, Parkinson's disease, Huntington's disease or CJD arise from excessive or premature death of specific classes of nerve cells. In each of these cases, mis-folding of proteins or fragments of proteins causes aggregation to form insoluble toxic aggregates that leads to death of a specific group of nerve cells. The quest to find effective treatments for these debilitating diseases includes attempts to prevent protein mis-folding, to re-dissolve insoluble aggregates of mis-folded proteins or to regenerate replacements for the damaged nerve cells by reintroducing neural stem cells into the brains of affected patients.

As improved healthcare extends the life expectancy of human populations, two of the major health challenges that remain and become increasingly important are neurodegenerative diseases and cancer. I have argued here that both of these major challenges to human health arise from imbalances in the life and death of cells.

References and further reading

Alberts, B., A. Johnson, J. Lewis, M. Raff, K. Roberts and P. Walter. 2008. *Molecular Biology of the Cell*, 5th edn. New York: Garland Science. Chapters 17 and 18.

Hanahan, D., and R. A. Weinberg. 2011. 'Hallmarks of Cancer: The Next Generation', *Cell* 144: 646–74.
Laskey, R. A. 2004. 'DNA and Cancer', in T. Krude, ed., *DNA: Changing Science and Society*. Cambridge: Cambridge University Press. 88–106.
Lodish, H., A. Berk, C. A. Kaiser, M. Krieger, M. P. Scott, A. Bretscher, H. Ploegh and P. Matsudaira. 2008. *Molecular Cell Biology*, 6th edn. New York: W. H. Freeman. Chapters 20 and 21.
Rubinsztein, D. C. 2006. 'The Roles of Intracellular Protein-Degradation Pathways in Neurodegeneration', *Nature* 443: 780–6.
Skloot, R. 2010. *The Immortal Life of Henrietta Lacks*. New York: Crown.
Vogelstein, B., and K. W. Kinzler, eds. 2002. *The Genetic Basis of Human Cancer*. New York: McGraw-Hill.

2 The spark of life

FRANCES ASHCROFT

Alex Mitchell achieved a curious kind of fame. He died while laughing uproariously at an episode of *The Goodies*, a famous British TV comedy show. The sketch featured a game of 'Ecky Thump', a spoof martial art contest in which opponents pelted one another with black puddings and defended themselves with a set of bagpipes. Alex found it hilarious and was convulsed with laughter throughout much of the episode. He let out a huge guffaw at one particularly amusing piece and then, to the surprise and consternation of his family, suddenly stopped laughing, collapsed on the sofa and died. The story of the 'man who died laughing' became headline news and his wife even subsequently wrote to the Goodies thanking them for making her husband's last moments so happy.

It was later found that Alex had a rare heart condition in which excitement can adversely affect the electrical activity of the heart, precipitating a cardiac arrest. Although it is perhaps not widely appreciated, humans are electrical machines. Everything that we think, feel and do is caused by electrical signals in our cells, from the beating of our hearts to our ability to see, hear, think, speak and move our limbs. We even define death as when the electrical activity of our brain ceases. Ultimately, this electrical activity and thus our thoughts, feelings, actions – even consciousness itself – is produced by a set of little-known but extremely important proteins called ion channels. This essay tells some of their remarkable stories and shows how Cambridge scientists played a crucial role in unravelling how the electrical signals in our cells generated.

The discovery of 'animal electricity'

The great Italian scientist Luigi Galvani was the first to discover 'animal electricity', in the late 1700s. Galvani was very interested in the recently discovered phenomenon of static electricity and experimented with the effects of electric sparks on living tissue. In a famous experiment, he discovered that an electric spark could cause the muscles of a recently dead frog's leg to twitch vigorously. As a consequence, he hypothesized that the electric spark was stimulating the muscle to contract. Galvani knew that lightning is also a form of electricity and thus he next investigated whether lightning could also cause the frog's legs to twitch. He attached the nerve that supplies the frog's leg muscle to a long metal wire and connected the wire to a metal spike, which he placed at the top of his house pointing towards the sky. To his delight, he found that the frog's legs did indeed dance around when lightning was about. A careful scientist, he next repeated the experiment on a calm day, as a control. This time, he hung the frog's legs from the iron railings of his balcony. To his surprise, he noticed that the frog's legs still twitched occasionally even when no lightning was around. Galvani took this to indicate that animal cells are not only stimulated by electricity, they actually produce their own – and he deduced that this 'animal electricity' was what caused the muscle to contract (Figure 2.1).

Galvani's idea was contested by his fellow countryman, Alessandro Volta who subsequently showed that the muscle twitches were in fact induced by an electric current that flowed between the iron railing and the brass hooks that Galvani had attached to the frog's nerve. These two different metals, he hypothesized, were generating an electric current – in effect they were acting as a simple battery. Volta went on to invent the first electric battery and Galvani's idea of animal electricity gradually fell into abeyance.

Initially, however, Galvani's experiment generated considerable excitement and gave rise to the idea that electricity was the 'spark of life'. Wherever frogs were available, scientists and laymen alike tried to reproduce his findings. Nor did they confine themselves to frogs – a wide range of other freshly killed creatures was employed. Most dramatic of all were the theatrical public demonstrations put on by Galvani's nephew Giovanni Aldini, who attempted to revitalize the dead. By applying electricity (in the form of one of Volta's batteries) to the corpse of a hanged criminal who had been quickly cut down and rushed to the nearby anatomy theatre, he induced fearsome facial grimaces, simulated breathing and contractions of

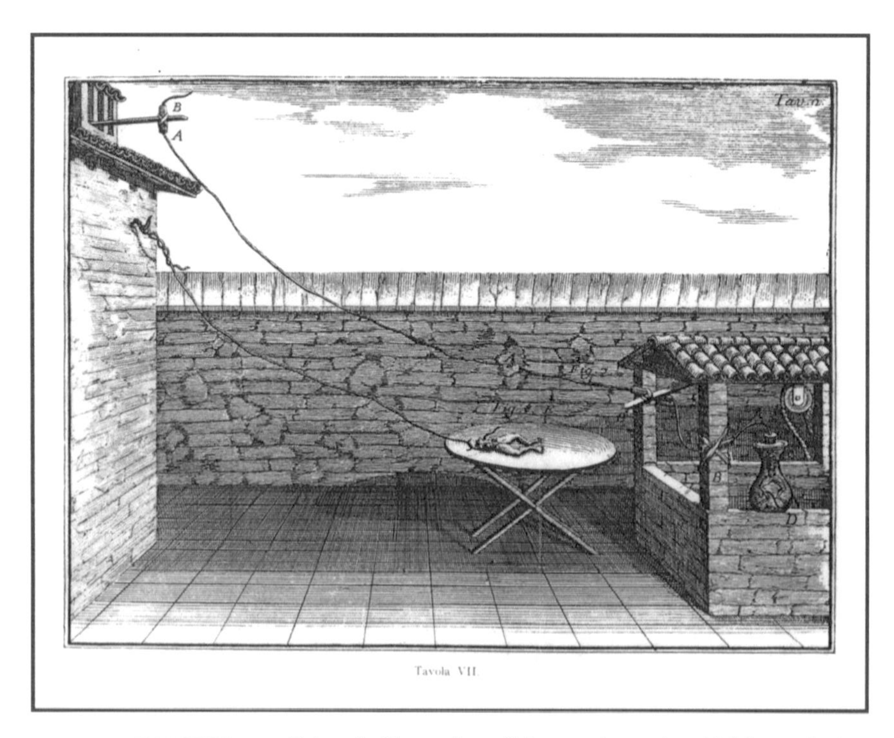

FIGURE 2.1 Galvani's illustration of his experiment in which he excited a frog's muscles with lightning. The frog's legs are seen lying on the table, with the nerve that supplies them connected to a wire that is hooked into the wall. A second preparation in seen in the flask on the far right, also connected to a spike on the wall on the left. All that is missing is the lightning itself.

the limb muscles. He wrote that, 'the jaw began to quiver, the adjoining muscles were horribly contorted, and the left eye actually opened'. A demonstration in Edinburgh produced even more appalling results, for the leg of the dead man kicked out, knocking over one of the assistants, and the arm lifted and appeared to 'point at the spectators, some of whom thought it had come to life'. Such spectacles may have been the inspiration for Mary Shelley's novel *Frankenstein* and the prevailing media image of the scientist as 'mad, bad and male'.

Unravelling the nerve impulse

Although Galvani's idea of animal electricity was initially discredited, it was evident to all that muscle contraction could be stimulated by an extrinsic electrical stimulus. And it was not long before it was shown that each muscle contraction is triggered by an electrical signal in the nerve that supplies it. These electrical signals are known as nerve impulse or action potentials. Because they are rapid, transient and very tiny it took many years for scientists to be able to measure them properly and to determine how they are generated and conducted along nerve fibres. Cambridge played a leading role in these studies.

One of the many eminent Cambridge scientists to work on the nerve impulse was Edgar Adrian, who started out as an assistant to Keith Lucas in the Physiological Laboratory just before the First World War. By painstakingly dissecting out a single nerve fibre he was able to show that when it was stimulated electrically, the fibre fired off a series of electrical impulses. The magnitude of each nerve impulse was always the same, independent of the stimulus strength, but a larger stimulus gave rise to a greater number of impulses. Information is therefore transmitted along nerve fibres in a form of digital code: each impulse is either 'all-or-none' and signal strength is coded by impulse frequency. Adrian won the Nobel Prize for his work, becoming the first of a long line of illustrious physiologists working on the nerve impulse to do so. He also became Master of Trinity College.

Adrian passed the baton to Alan Hodgkin and Andrew Huxley, two more remarkable Cambridge physiologists. They were the first to measure the nerve impulse – the change in electrical potential across the nerve cell membrane that is produced when a nerve is stimulated. Central to their success was the use of the squid giant nerve cell for their experiments. About 1 mm in diameter, it is large enough to be seen by the naked eye and to thread a wire electrode down inside it. This means one can record the voltage difference between the electrode inside the nerve cell and one placed in the external solution bathing the fibre. Hodgkin and Huxley found that there is a voltage difference across the nerve cell membrane at rest, the inside of the cell being about 70 mV more negative than the outside. When the nerve is stimulated, this voltage difference undergoes a rapid transient change, the inside becoming about 50 mV more positive than the outside for a brief period. This change in membrane potential constitutes the nerve impulse. It is also known as the action potential.

Like Adrian, Hodgkin and Huxley's research was interrupted by hostilities. Shortly after they recorded the first action potential, at the Marine Biological Laboratories in Plymouth in the summer of 1939, the Second World War broke out and they were seconded to war duties. It was a long eight years before they could return to their experiments. When they did so, they demonstrated that nerve impulse was the result of the transient changes in the flux of sodium and potassium ions across the cell membrane. In a series of classic experiments, they provided a comprehensive description of the electric currents produced by these changes in ion flow, and formulated a model of how the nerve impulse is generated and propagated that still holds today. They too were honoured with a Nobel Prize.

Hodgkin and Huxley, who were both young at the time of their prize-winning research, did not rest on their laurels, but went on to have distinguished careers in other areas of physiology. Huxley invented a new type of microscope and used it to solve another mystery – how muscle contracts. Hodgkin focused on the physiology of vision. Both became President of the Royal Society, the UK's national scientific academy, and (like Adrian) also Master of Trinity College.

Single-channel currents

We know now that the nerve impulse is due to the activity of ion channels. These proteins are tiny pores that sit in the membrane surrounding each of our cells. When the pore is shut, ions cannot pass through it, but when it opens, ions such as sodium (common salt) can permeate. It is the carefully timed opening and closing of two kinds of channel – those that selectively conduct sodium ions and those that allow only potassium ions to pass – that underlies the nerve impulse. Opening of the sodium channels triggers the impulse. Because sodium ions are higher outside the cell than inside it, when the channel opens they flood into the cell down their concentration gradient. And because sodium ions carry a positive charge, the inside of the cell becomes more positive, just as Hodgkin and Huxley discovered. After a brief delay, the potassium channels open. Because potassium ions are higher inside the cell than outside it, positively charged potassium ions move out of the cell through the open potassium channels,

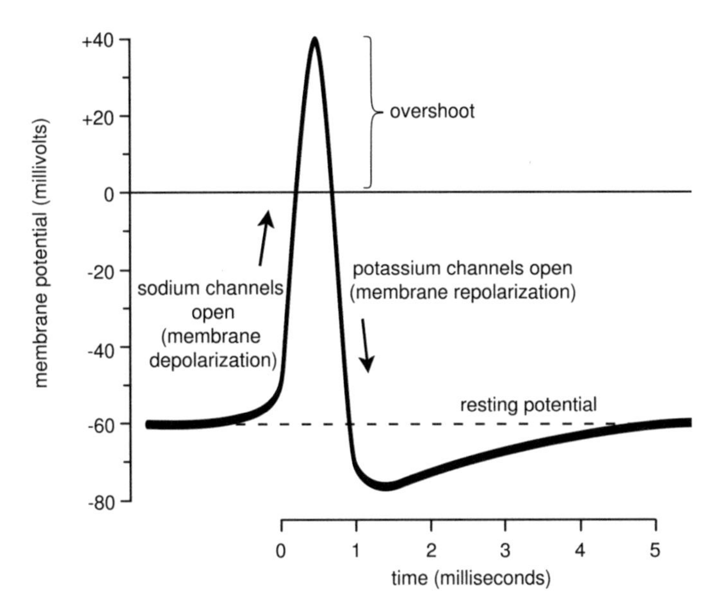

FIGURE 2.2 Action potential similar to that first recorded by Hodgkin and Huxley. The potential across the nerve cell membrane at rest is negative with respect to outside the cell, by around −60 mV. When the sodium channels open, the potential explosively becomes positive but then rapidly shifts negative again as the potassium channels open. The whole thing last for only about a millisecond.

restoring the charge difference. Consequently, the potential across the nerve cell membrane returns to the resting level. Subsequently, both types of channel close once more and the resting state is restored. The nerve impulse (action potential) is thus a transient change in potential produced as first the sodium channels open and then the potassium channels. This potential change sweeps along the nerve fibre from one end to the other, signalling information – for example, informing our brain that the fire is hot, or instructing our muscles to move our hand away from the heat (Figure 2.2).

Bacteria also have ion channels, including ones with giant pores that are permeable to many ions that are used as chemical weapons. These act like molecular hole-punches, being secreted by the attacking cell and inserting themselves into the membrane of the target cell where they make huge holes. Many different kinds of ions flow through the pore, as well as small molecules and water, causing the target cell to swell up and die. Some bacterial channel-forming proteins make very effective antibiotics. They have also advanced our understanding of how channels work: the very first recordings of current flow through a single ion channel were made by reconstituting a bacterial channel into an artificial membrane and measuring the current produced when it spontaneously opened. And the first quantitative description of such currents was another Cambridge achievement, reported by Steve Hladky and Denis Haydon, in 1970. To try to cut down vibrations that might destroy the artificial membrane and abort their experiment, they placed their equipment on a paving slab hung from the ceiling of the laboratory.

Even more excitement was generated six years later when the great German scientists, Erwin Neher and Bert Sakmann, succeeded in recording the current flowing through a single channel in a biological membrane. This was considerably smaller than the bacterial single-channel currents – just a few picoamps (a million millionth of an amp) – and it required the development of new equipment, sensitive enough to be able to pick up the tiny signals. In those days, physiologists needed to have an excellent knowledge of physics and electronics, and to be capable of building their equipment themselves. Neher and Sakmann added yet another Nobel Prize to the tally garnered by ion channels and the nerve impulse.

Toxins – for good and ill

Fascinatingly, ion channels are the target for many toxins. Evolution discovered that blocking their function, and thus nerve and muscle impulses, is a good way to incapacitate a predator or immobilize one's prey. A vast array of poisons from spiders, shellfish, sea anemones, frogs, snakes, scorpions and other creatures interact with ion channels and block their function. Some of these neurotoxins act very rapidly. A cone snail can kill a fish in just a few seconds by injecting it with a cocktail of around 200 different toxins, several of which act by blocking sodium and calcium channels.

One of the most famous neurotoxins is tetrodotoxin, which is found in the liver, intestines, skin and ovaries of the puffer fish. A virulent poison, it inserts itself into the outer mouth of the sodium channel, plugging the pathway through which the sodium ions move. This prevents the transmission of electrical impulses in nerve and muscle fibres and so produces paralysis. The victim dies of asphyxiation because their respiratory muscles do not work. There is no antidote, but if the person is artificially ventilated they can be kept alive until the poison has washed out of their system. In Japan, puffer fish is known as *fugu* and is served in top restaurants as sashimi – thin slices of raw muscle. It is a highly prized and very expensive delicacy but unless it is carefully prepared the flesh can be toxic and although all *fugu* chefs must be specially trained and licensed, several people die each year from tetrodotoxin poisoning (usually fishermen who have eaten their own catch).

Not all toxins work by blocking sodium channels – some produce equally devastating effects by locking them open. One of these is grayanotoxin, which is found in honey made from the nectar of some varieties of rhododendron. The Greek soldier Xenophon recounts how his men were incapacitated after consuming 'mad honey' on the shores of the Black Sea in Turkey – the symptoms included weakness, vomiting, heart arrhythmias and even convulsions. Three hundred years later when King Mithridates and his troops were under attack by the Roman general Pompey, they strewed mad honey in the path of the enemy, who ate it avidly, were rendered *hors de combat* and thus easily massacred by the Greeks. It was an early form of chemical warfare.

But agents that interfere with ion channel function are not always to be feared. Used in the right way, to block the right channel, they can be very important therapeutic drugs. The local anaesthetic that numbs the pain of dental operations, for example, works by blocking sodium channels. Unfortunately, it does not discriminate between sodium channels in pain nerve fibres and those in the motor nerves that innervate the facial muscles so although you will feel no pain you also often end up with a lumpen jaw. However, because sodium channels in pain fibres and motor nerve fibres are not identical, it is hoped that it will be ultimately possible to develop drugs that target only the former.

Mind the gap

Whenever we decide to move a muscle, our brain sends an electrical impulse out along the motor nerve fibre to its endings on the muscle.

However, the electrical signal cannot jump the gap that separates a nerve and muscle cell – instead a chemical transmitter is used. At the nerve–muscle junction the chemical is called acetylcholine and it is packaged into tiny membrane-bound vesicles that are found at the tip of the nerve endings. The arrival of the electrical impulse triggers these vesicles to fuse with the surface membrane and dump their contents into the gap between the two cells. The transmitter then diffuses across the gap, where it binds to channels known as acetylcholine receptors in the muscle membrane, and triggers an electrical impulse in the muscle that causes it to twitch.

A whole range of drugs and toxins act on ion channels at the nerve–muscle junction. One of the best known is curare, the poison that South American Indians use to tip the darts they use in their blowpipes. Curare blocks the acetylcholine channels in the muscle membrane and so prevents the nerve from stimulating the muscle fibre. Consequently, the animal is completely paralysed, falls from the tree and dies from respiratory failure. Drugs based on curare are widely used in operations to prevent muscle movements that might hamper the surgeon's work: the patient is artificially ventilated to help them breathe.

Clearly, if a muscle is to be able to respond to a second nerve impulse, the first signal must be switched off rapidly. This is achieved by an enzyme called acetylcholinesterase, which destroys acetylcholine. Agents that inhibit the action of acetylcholinesterase are lethal because they lead to the build-up of acetylcholine in the tiny gap between the nerve and muscle fibres, and thus to overstimulation of the acetylcholine receptors and, as a result, muscle convulsions. The nerve gas sarin works this way. In 1995, the Aum Shinrikyo sect released sarin in the Tokyo Metro. Twelve people were killed, fifty seriously injured and the sight of almost a thousand others was temporarily impaired.

The heart of the matter

The contraction of the heart is triggered by electrical impulses in the muscle fibres of the heart. These signals originate in the pacemaker cells of the sinus node, which sits in the upper right chamber of the heart, and they spread across the heart via a network of specialized cells so that they reach every cardiac muscle fibre. The timing and spread of the electrical impulses is co-ordinated to ensure that the left and right chambers of the heart contract in synchrony, the upper ones first and then the lower, larger chambers, which pump the blood around the body. If the electrical signals are distorted, the heart will no longer beat rhythmically and blood flow will be disrupted, leading to brain damage and eventual failure of the heart to contract.

Several different kinds of ion channels contribute to the electrical signals that control the beating of our hearts and mutations (defects) in the genes that code for these channels can give rise to a relatively rare inherited cardiac disorder that causes abrupt loss of consciousness and sudden death. Known as Long QT syndrome, it is a particularly tragic disease because people often die when they are relatively young.

The strange name of this disease comes from its effect on the electrocardiogram (ECG), the electrical signature of the heart. As the heart beats, electrical signals (the action potentials) in the heart cells give rise to tiny fluctuations in electrical potential on the surface of the body. These can picked up by the surface electrodes on the chest and constitute the ECG. It provides a very good indication of what the individual heart cells are actually doing. The upstroke of the cardiac action potential is associated with the Q wave of the ECG and the down stroke with the T wave. Thus, the longer the duration of the action potential, the greater the interval between the Q and T waves of the ECG. Some people have mutations in their cardiac ion channels that cause an increase in the duration of the cardiac action potential and thus their QT interval becomes longer – hence, Long QT syndrome. A long QT interval can precipitate electrical arrhythmias in the heart, which can be fatal because the heart no longer beats in a co-ordinated fashion and thus is unable to pump properly. Instead, electrical chaos means the individual cells beat asynchronously – it looks, the great anatomist Vesalius once remarked, 'like a quivering bag of worms'. When this happens the heart rhythm must be reset by first stopping it and then restarting it using a defibrillator.

In people who carry long QT mutations, cardiac arrhythmias are often triggered by exertion, stress or excitement. In the case of Alex Mitchell, it was the excitement of watching the riotous antics of *The Goodies* that did it – he really did die of laughing. But it was not until many years later that his condition was recognized, when his granddaughter suffered a cardiac arrest at the age of twenty-three. She was saved by her husband who carried out cardiopulmonary resuscitation until the ambulance arrived. It was then discovered that she – and her grandfather – carried a mutation that caused Long QT syndrome. Fortunately, there are now drugs that can be used to treat Long QT syndrome and allow patients to have a normal life. They may also be given an implantable defibrillator, a device that detects when the heart beats arrhythmically and then delivers an electric shock to stop the arrhythmia. Often when the heart spontaneously restarts it beats normally (it is rather like rebooting a computer and finding the problem has vanished when it restarts).

Long QT syndrome can also be induced by drugs that block cardiac ion channels. The drug Terfenidine is a very effective anti-allergy agent and used to be sold over the counter in the UK. It was changed to being a prescription-only drug a few years ago when it was discovered that in a few people it could cause Long QT syndrome. All novel drugs must therefore now be screened for their effects on cardiac ion channels to ensure that they do not induce Long QT.

Myotonic goats come to Cambridge

Mutations in skeletal muscle ion channels can also cause disease. One that has an interesting Cambridge connection is an inherited disorder of muscle stiffness known as myotonia congenita. It was first described, in 1876, by Julius Thomsen, a Danish physician, in himself and his family. He hid his complaint until he was in his sixties, when one of his sons, who also suffered from the disorder, was accused of malingering and using his condition to avoid military service. Thomsen published his studies to defend his son.

People with myotonia congenita are unable to relax their muscles easily; anything that involves a strong contraction tends to result in muscle 'cramp'. Pick up a heavy suitcase, for example, and they may find they are unable to release their grip when they put it down. Try to do something quickly and they find their muscles seize up. One man said when he crouched down to start a race, his legs would lock into complete extension when the starting gun went off. 'It was like trying to run on stilts.' He only actually start running properly after he had covered some ground and his muscles had 'warmed up'. Myotonia congenita is not only inconvenient; it can also be dangerous. When a tree accidently fell towards him, a lumberjack with the disease found himself frozen to the spot, unable to move. He was lucky, for the tree knocked him backwards into a hollow in the ground but did not crush him.

A similar disorder is found in a rare breed of goats, known as myotonic goats, stiff-legged goats or fainting goats (the latter name is misleading as the goats do not actually faint). Originally, these animals were bred for their meat – and they were easy to keep as every time they tried to leap over the fence out of the field they would simply topple over. The reason they are favoured as meat animals is that their muscles are constantly undergoing tiny contractions, so they tend to be very well developed. It is as if they are performing continuous isometric exercises. Nowadays, they are more usually kept as pets, for their novelty value.

The goats turned out to be the key to understanding the human disease. In the 1960s Shirley Bryant of Cincinnati serendipitously read about them in a veterinary journal. He was intrigued by stories of the whole herd simultaneously falling over when the train sped past their field and sounded its whistle and, realizing that the condition had similarities to human myotonia, he decided to investigate. Having acquired his own herd of myotonic goats, he showed that a single stimulus produces a single electrical impulse in a normal muscle fibre, but that in myotonic muscle it produces a burst of impulses that may continue even after the stimulus has stopped. This explains why the myotonic muscles contracted more – they were simply being excited more.

Bryant was eager to unravel the origin of this hyperexcitability. His work suggested that the problem lay in the movement of chloride ions across the muscle membrane. Chloride fluxes play an important role in muscle, damping down excitability, and Bryant postulated that chloride flux was impaired in myotonic muscle. To prove this he needed to be able to record the electric currents carried by chloride ions as they move into and out of the muscle cell. That required specialist equipment, which he did not have. Fortunately, it was available in Richard Adrian's laboratory in the Physiology Laboratory at Cambridge. Bryant therefore decided to bring his goats to the UK.

This was less simple than it sounds. Due to fears that the animals might carry a disease called bluetongue, a special act of Parliament had to be passed to allow their importation. Bryant travelled to London in advance of the goats, to ensure that all was ready for their arrival. He relied on a colleague to send off the animals and, crucially, to ensure that all the necessary paperwork was sent with them. Alas, his colleague forgot to include the documents, and the goats arrived at Heathrow airport without them. They were immediately impounded and in imminent danger of being shot. Adrian, alerted by a frantic phone call from Bryant, summoned all his powers of persuasion, argument and charm (which must have been quite considerable) to convince the authorities that the dictate to slaughter all animals that arrived without papers applied to those which were 'disembarking' (from a ship) and not those that were 'disemplaning'. He obtained a day's stay of execution by which time the necessary papers had fortunately arrived.

It turns out that, as Bryant had hypothesized, goats – and humans – with myotonia congenita have a mutation in the gene that codes for their muscle chloride channel that renders its non-functional. As a consequence, the muscle becomes hyperexcitable.

Ion channels are not only found in our nerves and muscles. They are present in each and every one of our cells – indeed, in all cells on Earth – and they regulate every aspect of our lives. One of the most complex channels in the human genome is known as CatSper. It is found only in sperm and it plays a special part in our lives, as it is crucial for winning the great sperm race. Strange as it may seem, your own existence is dependent on the activity of this ion channel in the single sperm cell that fertilized the egg from which you developed.

Sperm must swim from the moment of ejaculation, fighting their way towards the egg by lashing their tails. For much of the way up the female reproduction tract they simply wriggle sedately along but in the vicinity of the egg, they become hyperactivated and switch to larger and more forceful whips. This last-minute turbo charge is essential because it provides the power the sperm needs to penetrate the membranes surrounding the egg. It is triggered by the opening of CatSper channels, which are located in the tail of the sperm. If this change in beat fails to happen, the sperm cannot fertilize the egg. Each CatSper channel is made up of a number of sub-units encoded by different genes, and infertility is produced if any one of them fails to function properly. As the CatSper channel is unique to sperm, drugs that prevent its channels opening might make a good contraceptive.

Smell, taste, sight, sound, touch – all sensation involves the conversion of an external stimulus into an electrical signal that can be received, acted upon and memorized by the brain. And that electrical signal, in turn, is due to the activity of specialized ion channels. Ion channels sensitive to mechanical vibrations sit in the hair cells in our ears and convert the pressure waves in the fluid of the inner ear into sounds that we can hear. A related channel in the cone cells of the retina (the layer of photosensitive cells at the back of the eye) enables us to see colour. Because cones are also important for visual acuity, people with mutations in this channel see a fuzzy, monotone world. Wonder why chilli peppers taste hot? It is because the active ingredient (known as capsaicin) stimulates the same ion channels as those involved in the detection of noxious heat. The brain cannot tell the difference and interprets both as heat. A variety of this type of channel has even been co-opted as a heat sensor by vampire bats, to enable them to detect their warm-blooded prey. They are truly the doors of perception.

Any Natural Science student at Cambridge cannot fail to be aware of its proud tradition of research in the area of nerve and muscle electrical impulses and I was no exception. Eventually, I, too, ended up studying electrical signals and ion channels, although my main interest is their role in the regulation of insulin secretion. Insulin plays an essential role in controlling the blood sugar concentration. Too high or too low a level of sugar (glucose) in the blood are both bad for you. If blood glucose falls too low, even for a few minutes, the brain is starved of fuel, leading to irreversible brain damage. Conversely, too much sugar, for too long, results in the complications of diabetes mellitus – heart disease, kidney disease and blindness, to name just a few. Insulin is secreted from the beta-cells of the islets of Langerhans (small islands of cells that lie within the pancreas) in response to a rise in the blood sugar concentration. It lowers blood glucose levels by promoting uptake of the sugar into muscle, liver and fat cells and it is the only hormone capable of doing so, which is why insufficient insulin leads to diabetes.

Our studies showed that a potassium channel plays a central role in glucose-stimulated insulin secretion. In the absence of glucose this channel is open, which inhibits beta-cell electrical activity, and, in turn, prevents insulin secretion. When the beta-cell is exposed to glucose, however, the channel closes, stimulating electrical activity and insulin release. In essence, the channel acts as molecular switch that links the metabolism (breakdown) of glucose by the beta-cell to insulin release. Another Cambridge scientist, Nick Hales, was the first to show that a chemical known as ATP, produced by glucose metabolism, binds to the channel and causes it to close. Consequently, the glucose-sensitive channel is now known as the K_{ATP} channel (K is the scientific symbol for potassium). Hales' team was also the first to show that drugs known as sulphonylureas stimulate insulin secretion by shutting the K_{ATP} channel. Despite the fact drugs had been used very effectively for more than thirty-nine years to treat Type 2 diabetes, up until that time their definitive mode of action was unknown.

More recently, it has been discovered (by Andrew Hattersley and Anna Gloyn at Exeter University) that mutations in the genes that code for the K_{ATP} channel sub-units can cause a rare form of diabetes, known as neonatal diabetes, that usually manifests within the first six months of life. We were able to show that these mutations prevent the K_{ATP} channel from closing in response to a rise in blood sugar, so that although the beta-cell makes insulin as normal it is unable to release it. This has had a happy outcome for patients with neonatal diabetes. In the past, their diabetes was treated with insulin injections but our joint work with the Hattersley team showed that they should be able to take sulphonylurea drugs instead. Several hundred patients have now made the switch. What allowed this to happen so swiftly was both an understanding of exactly how the K_{ATP} channel acts to regulate insulin secretion and the existence of a drug that was in routine clinical use for treating diabetes, which could be immediately used in another patient population. Of course, to obtain this knowledge took many, many years of research by academic scientists, clinicians, and the pharmaceutical industry!

This essay has provided a brief overview of a selection of stories about ion channels and electrical signals in the cells of our bodies. It has had a deliberate Cambridge focus. Many fascinating stories have been left out, such as the importance of ion channels in fighting infection, regulating blood pressure, learning and memory, and their role in diseases such as cystic fibrosis, startle disease and epilepsy. Nevertheless, I hope that I have convinced you that ion channels are an exciting field of study and that they are indeed the spark of life.

3 From genomes to the diversity of life

MICHAEL AKAM*

Introduction

A genome is the complete set of genetic instructions for an organism. For each of us, it is forty-six large molecules of DNA in the nucleus of every cell, each packaged into its own chromosome, together with many copies of a shorter fragment of DNA that lies in a specific compartment of the cytoplasm.

This genome makes us what we are, by determining, directly or indirectly, the structure of every complex molecule in our body. Together with a little help from the genomes in our parents, and our parents' parents, it determines how our bodies are put together, and how they function, and how they interact with the myriad environmental influences before and after birth to make us what we are today.

Genomes are by no means the only, or even necessarily the best, level of organization at which to study how all these complex things happen, but they are interesting. In this chapter, I look at some of what we have learnt about life by studying genomes.

Gene and genome sequencing

The recent history of genome sequencing has been one of the most remarkable technology success stories.

It is nearly forty years since the invention of fast techniques to sequence DNA, by Fred Sanger at the Laboratory of Molecular Biology (LMB) here

* I have drawn on the work of many colleagues in this discussion. I thank particularly Detlev Arendt, Rob Asher, Carlo Brena, Eric Davidson, Peter Holland, Nicole King, Chris Lowe, Mark Martindale, Rudy Raff and Giselle Walker, who have in one way or another particularly influenced what I have presented in this chapter.

in Cambridge, and Walter Gilbert's laboratory in the US. In 1979, I spent some months working as part of a team at the LMB, and contributed to a project which by 1982 had sequenced the entire 6395-base genome of tobacco mosaic virus. We were proud of that.

By the end of 2010, the largest sequencing centre in the world, the BGI at ShenZhen in China, could generate more than five terabases of raw sequence data per day – that is 1000 human genomes.

We are all familiar with developments in computer power over the last thirty years – a trend described by Moore's law, which says that computer power doubles every two years. In recent years, sequencing power has done very much better than this. Since about 2007, when next-generation sequencing technologies became available, sequencing costs have plummeted from about US$10 million for a human-sized genome, to less than US$10,000 – and with confident predictions of the US$2000 genome, completed in two hours, by 2013.

In fact, in the last couple of years, the ability to sequence DNA has ceased to be a rate-limiting step for most biological analyses. Analysing this data has become the limiting step. Just assembling the short lengths of sequence that come out of the sequencing machines into meaningful fragments of chromosomes requires extraordinary computer power – BGI has a computer with 200 terabytes of RAM. But once that is done, the real scientific challenges begin: how do we interpret all of this DNA sequence – what can we read from genomes?

Here I look at just a few aspects of that question: what it tells us about the history of life, and the processes of evolution that have generated the diversity of life around us today. I am not concerned with impacts on human health, or our understanding of human history – though if I were starting my career again I think I might well choose genetic archaeology as the most exciting of topics to work on.

I must confess at the outset to a dreadful bias. I am going to ignore almost every form of life that does not have a nucleated cell – that is, all of the bacteria, the archaea and the viruses, and the origin of life itself. I am going to focus on that branch of the tree of life that began with the coming together of several different ancestral organisms, each with a genome of its own, to form a new type of unicellular life, at least 2000 million years ago. This new type of cell gave rise to all of the animals, plants and fungi alive today, as well as a myriad of other forms. We call all of these organisms 'eukaryotes' or the 'truly nucleated' cells. I am going to talk almost exclusively about eukaryote genomes.

The first eukaryote genome to be sequenced was that of brewer's yeast, *Saccharomyces cerevisiae*, in 1996. Since then, hundreds of eukaryote genomes have been sequenced, and there are many more in the pipeline. At first, these genomes were all popular lab 'model organisms', but increasingly, all branches of the diversity of life are being sequenced.

Genomes and the tree of life

From Darwin's very first sketches of trees of relatedness of animals, it did not take very long before biologists were beginning to draw really rather complicated trees depicting how all organisms might be related to one another. Ernst Haeckel drew one of the most famous (Figure 3.1). Like most of the biological trees that were produced until the later part of the last century, it combined a mixture of real insight with speculation and prejudice. For example, man is right at the top of Haeckel's tree, and, by and large, as you go up the tree, things become more complex. Near the bottom of the tree, one small twig represents something called Infusoria, the single-celled organisms that appear in ponds and swim around in the water. As you will see, our perspective on the diversity of life has changed quite a lot since this tree was published.

A genome sequence is not only a coded set of instructions, but it is also a record of history. There are many types of changes that happen to the genome. Some are changes of single bases, from a C to a T, for example, which have relatively little significance for function, but which simply represent noise in the process of copying DNA. Changes like these accumulate over time, recording the history of genomes. If genomes share many of these changes, they presumably come from closely related organisms. There are other more complex changes which record history, for example in the order of genes on chromosomes. So genome sequences provide us with a way of looking at the history of life.

It was in 1988 that the first major paper was published purporting to tell us the relationships of the animals based on gene sequence data.

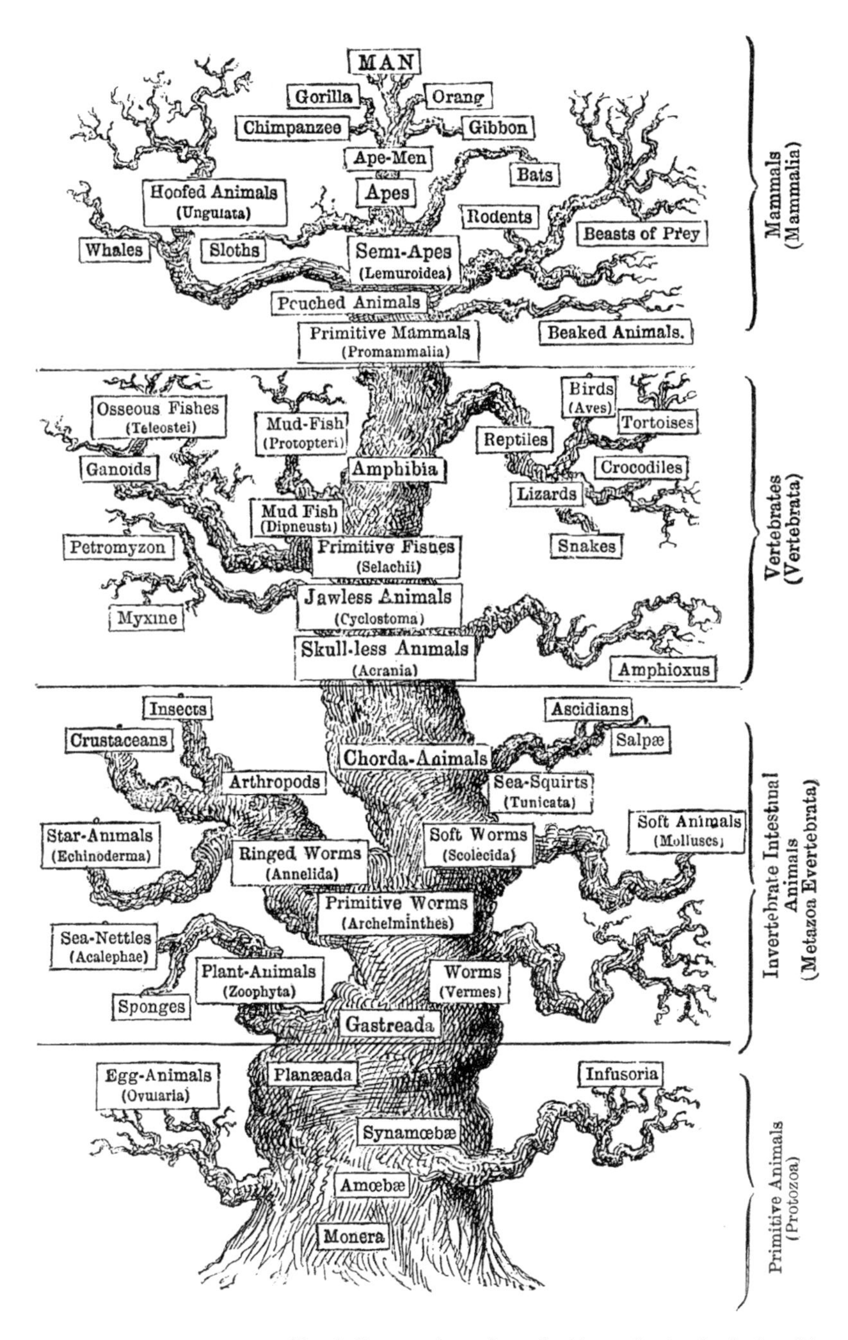

FIGURE 3.1 Haeckel's tree shows how far ideas of animal relationships had crystallized within twenty years of the publication of Darwin's theory. Note that man is firmly placed at the top of the tree. (Figure reproduced from Ernst Haeckel, *The Evolution of Man*. London: Kegan Paul, 1879.)

This paper, modestly entitled 'The Phylogeny of the Animal Kingdom', was based on the sequence of one gene that is present in all eukaryotes, the gene encoding a ribosomal RNA molecule, about 2000 bases long. Because of developments in sequencing techniques it had become possible to read the sequence of this gene in a relatively large number of different species – representing twenty-two different classes of animals in ten of the different major groups, or Phyla. It was published in the high-profile journal *Science*, and it made a big splash. I remember being very excited by it.

What is most striking about this paper is that almost all of the major conclusions that were highlighted in the abstract are now regarded as wrong. For example, it claimed that the multicellular animals (technically, the Eumetazoa), had two independent origins from single-celled ancestors – that the sea anemones and their radially symmetric relatives came from one lineage of protists, while the bilaterally symmetric animals came from another. A new and radical conclusion, almost certainly wrong.

So this was an approach to biology that had its teething problems, but in the twenty-five years since then, our ability to understand the processes that generate historical changes in genomes, and to infer from those changes the pattern of relationships among animals, has improved enormously. Importantly, recent analyses are based not on a single gene but on the sequences of a significant fraction of the genes in whole genomes. These new data are providing us with a pattern of relationship that is much more robust and, in some respects, almost as surprising as those earlier erroneous conclusions.

The first thing I would like to highlight is our place, as multicelled animals, in the bush of life (Figure 3.2). We are not at the top of the tree. It turns out that the major diversity of eukaryote life lies among those Infusoria – the single-celled animals now called protists. The many different lineages of single-celled nucleated organisms have been evolving for perhaps 2000 million years, and they are extraordinarily diverse in every respect – molecular biology, structure and genome sequence. The relationships of all these different single-celled forms, and of the multicellular forms that came from them, are only now becoming clear, since we have been able to sequence the genomes of many.

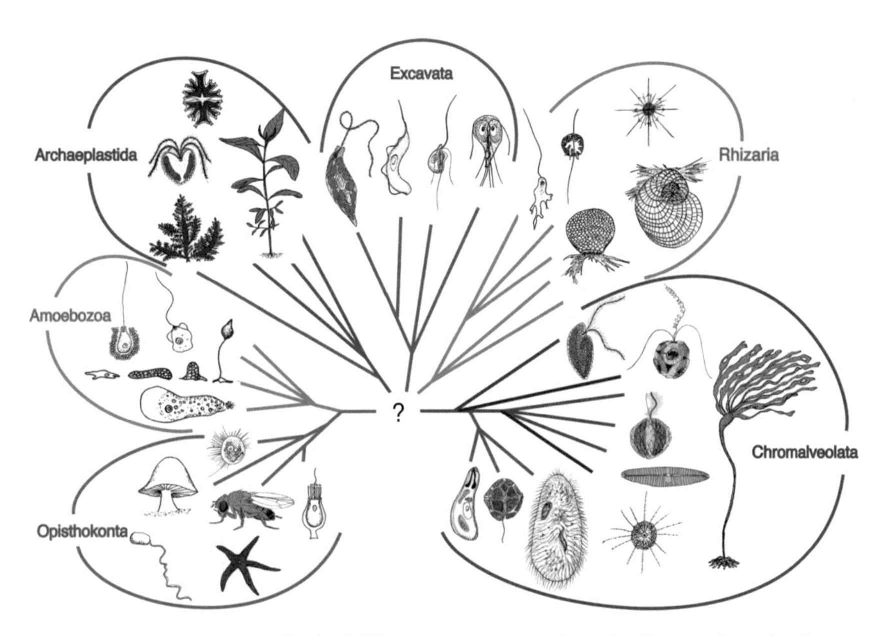

FIGURE 3.2 In the billion or more years since the first nucleated cells arose, their descendants, termed eukaryotes, have adopted a great variety of forms. We now recognize that the entire diversity of the animals (here represented by a fly and a starfish) represents only one twig on this bush of life. In terms of genomic structure and cellular function, the array of unicellular forms, termed protists, are much more diverse. The major multicellular lineages – animals, plants, fungi and the seaweeds – have arisen from different unicellular ancestors. (Picture courtesy of Giselle Walker.)

We, the multicellular animals, represented by a fly and a starfish, are relegated to a single twig at the bottom-left-hand corner of this bush of life. In fact, ten years ago, when I began to teach lectures on animal diversity, I had no idea where in that bush the genetic lineage of animals arose. There were suggestions that they came from ciliates, or from amoebae, and many other possibilities. But we now have a very good idea of the lineage of single-celled creatures that are our closest relatives among those protists. And it turns out that a very old idea was correct.

In the 1880s, William Saville-Kent described one lineage of those single-celled organisms, the choanoflagellates, which are widespread as plankton in the sea. The cells of these rather delightful creatures have a single beating flagellum that causes a current of water to flow through a

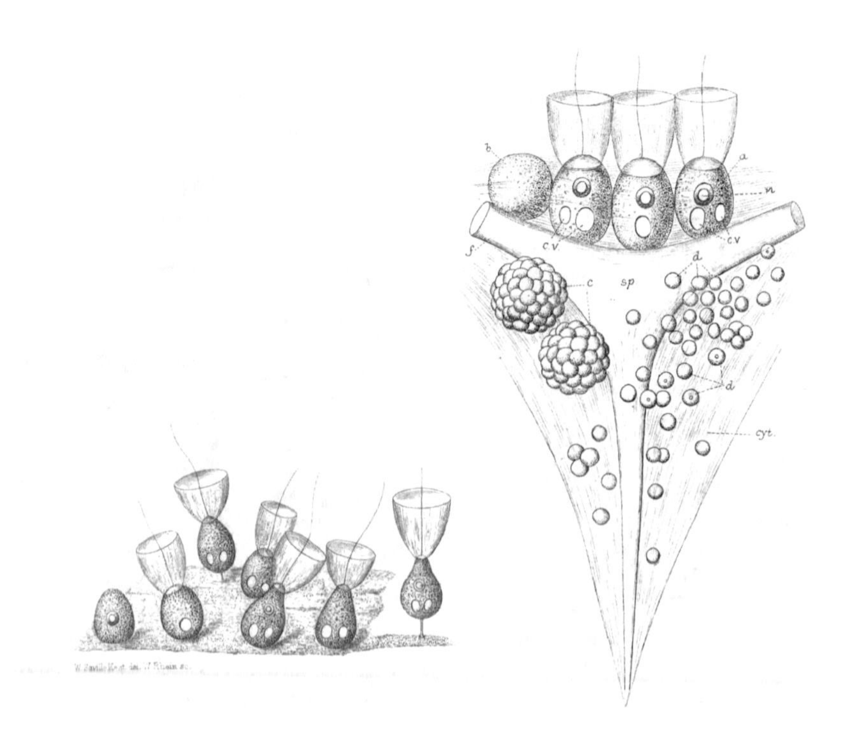

FIGURE 3.3 Data from genome sequences have identified the choanoflagellates as the closest living unicellular relative of the animals. This confirms an idea first put forward by Saville-Kent who, in the 1880s, noted the striking similarity between choanoflagellates, bottom left in the figure, and the filtering cells of the sponges, termed choanocytes (right). We now recognize sponges as the most basal group among the animals. (Figures taken from Saville-Kent's *Manual of the Infusoria*. London: David Brogue, 1880–1882.)

collar of very fine tentacles that surround its base (Figure 3.3). These filter out bacteria and other particles upon which the cell feeds. Saville-Kent noticed that these choanoflagellate filtering cells were extraordinarily similar to the filtering cells in sponges. He suggested that choanoflagellates and sponges were closely related.

Sponges were recognized long ago as sharing many characteristics with other animals, and are now confirmed as one of the earliest diverging branches in the tree of animal life. For about 100 years, though, the hypothesis that animals arose from a choanoflagellate-like form was simply one of many competing ideas. However, with the first cloning of genes from choanoflagellates, hints began to accumulate that this may be true.

In 2008, the genome of one choanoflagellate was sequenced, an organism called *Monosiga brevis.* This made it very clear that choanoflagellates share many genes uniquely with the rest of the animals – genes that are not found anywhere else in the great diversity of unicellular forms. They are our closest known living relative among the protists, and so we guess that the last unicellular ancestor of all animals might have looked something like a modern choanoflagellate. Indeed, it turns out that choanoflagellates are rather good at switching from being single cells to living as a small group of cells. Under certain feeding conditions they stick together and swim around in clumps. This might be a model for the lifestyle that first gave rise to multicellularity among the animals. The genome of choanoflagellates has now become an interesting place to look for the genetic origins of our own multicellularity.

I now go right to the opposite end of the phylogenetic spectrum, to illustrate another aspect of the diversity of animals, where genome sequences have helped us. Consider three mammals: a European mole, the sort of animal that might pop up in the middle of a lawn in Britain; a Cape golden mole, which might pop up in the middle of the garden if you lived in South Africa; and an elephant. It might seem obvious that the two moles are most closely related, but genomes tell us that this is not true!

I am cheating a bit here, because it has been known for a very long time that golden moles are not actually close relatives of our European moles. The traditional discipline of comparative morphology, studying the details of their skulls and other structures, had long shown that these two groups of moles must have arisen from different groups of mammals. Selection has been a powerful force in restructuring their bodies, making animals with similar lifestyles look the same. This convergence has made it extremely difficult to figure out where they should go in the tree of the mammals.

In fact it is the golden mole and the elephant that are the closer relatives. Sequencing the genome of mammals in the last ten to fifteen years has revealed that there is a diverse group of African species that share a large number of genomic characteristics which identify them as being more closely related to one another than they are to any other animals. These include tenrecs, aardvarks, elephant shrews, dugongs and manatees, as well as golden moles and elephants. These are now known as the 'afrotheria'. Many of these are species that had been hard to place, but before the molecular sequences began to arrive, I do not think that anybody had put all of them together in the same group.

Let me suggest one example of the evidence on which this conclusion is based. In our genomes, as well as the useful genes that encode parts of our own structure, there are also sequences that are there largely for their own purposes, principally to make extra copies of themselves in the genome. These 'selfish genes' can jump from site to site in the genome and insert into new positions. When a gene jumps into a new position it is quite likely to stay there, providing a signature in the DNA linking all the descendants of the ancestor in which the jump happened. The afrotherian mammals share several such insertions, and other rare gene rearrangements, uniquely to the exclusions of other animals.

Results like this suggest that geography has actually been much more important in the evolution of mammals than was previously recognized. The dramatic radiation and diversification of one early lineage of mammals in Africa made it very difficult to see what their ancestry was, but there is sufficient information in the genome to tease this history apart.

What's in a genome?

The most obvious things that we expect to find in a genome are . . . genes.

For a molecular biologist, a gene is most simply thought of as that portion of a chromosome that contains the coded message to direct the synthesis of a protein – a protein which might be part of the structure of a muscle, or a hair, or it might itself be a micromachine, an enzyme, to copy DNA, or digest food, or whatever.

This appears rather different from the original conception of a gene defined by breeders – Gregor Mendel's gene for round or wrinkled peas, for example. We now realize that round and wrinkled peas are the result of carrying slightly different versions of the code for one particular protein. All peas have essentially the same set of genes, but they have different versions (alleles is the technical term) for many of these genes, and it is these differences that make different pea plants different.

One of the first questions that everyone asks of a newly sequenced genome is 'How many genes does it contain?' That turns out to be not so easy to answer.

In the 1970s and 1980s, the best estimate for the number of genes in a simple animal like a fruit fly was about 5000. This was based on a combination of classical breeding experiments and looking at bands on chromosomes.

It was generally assumed, I believe, that a complex organism like man would have far more genes – 100,000 was widely suggested as a plausible number. That was a comfortable feeling, that man had twenty times as many genes as a fly. I certainly believed this, as late as 1992. In the venerable tradition of betting on the outcome of experiments, Martin Evans and I entered a bet on the number of genes in a mouse. Martin, now Nobel Laureate famed for the discovery of mouse embryonic stem cells, and then a colleague in the Wellcome/CRC institute here in Cambridge, thought that the number would be smaller than 100,000. Martin was right – by any plausible definition of gene, a mouse has fewer than 100,000 of them.

By 2002 it was clear that both the estimate for the fly and the estimate for man were wrong, but in opposite directions. Single-celled bacteria and single-celled 'higher organisms' like yeasts have about 5000 genes, but animals like flies and worms have more – often about 15–20,000. At that time, mammals – mouse and human, for example, were thought to have about twice the number – 30–40,000. Even that modest superiority to the flies has been revised again – recent estimates suggest that the number of genes coding for protein in the human genome may be a mere 20,000 or so. But, in fact, it is not so easy to say how many genes there are in any genome.

The easiest genes to recognize are those that code for proteins. We know the code that translates DNA into protein sequences, so finding such sequences should be easy. Indeed, for many genes it is, even though coding sequences are often interrupted by stretches of non-coding DNA that get spliced out of the final message. Automatic genome annotation

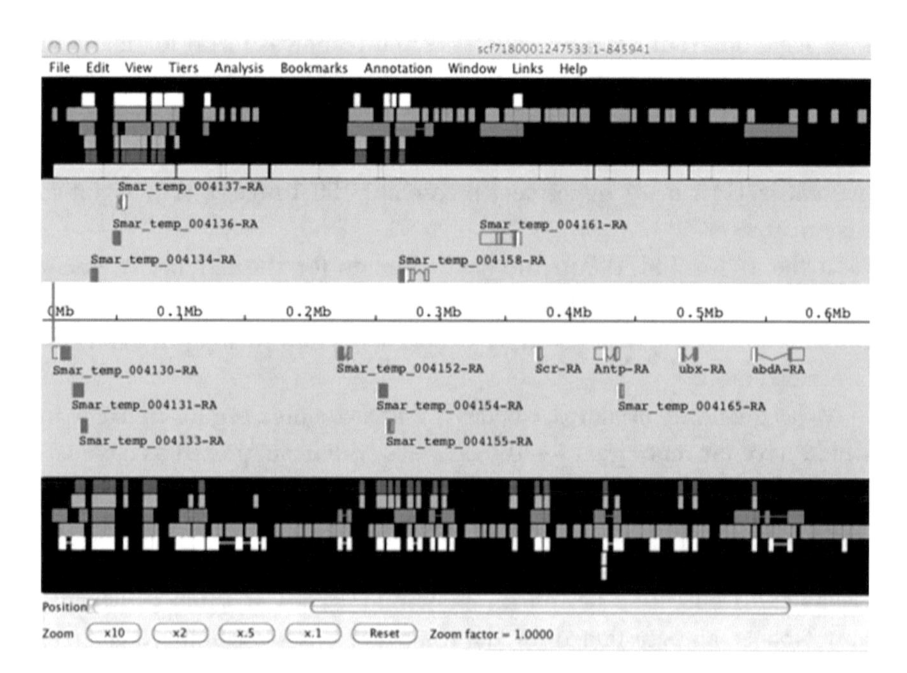

FIGURE 3.4 Even the most interesting parts of the genome are meaningless when viewed simply as a string of As, Cs, Gs and Ts. To make this sequence interpretable, annotation programs use knowledge of how DNA is 'read' by the cell, together with experimental data, to predict where genes are located. Genome browsers display this information in an intelligible way. The screen shot taken here, from the genome browser 'Apollo', shows predicted genes in a 640,000-base pair region of the Hox gene cluster of my favourite centipede. Above and below the DNA scale in the middle are boxes showing the predicted genes on the forward and reverse strands of the DNA, and how they may be copied and spliced into RNA messages. Bars at the top and the bottom of the view show different aspects of the evidence used to generate the predictions. These may include open reading frames in the sequence, similarity to other known genes, and regions with similarity to sequenced RNA transcripts.

programs can find perhaps the first 15,000 or so genes in a typical animal genome with little difficulty (Figure 3.4).

But then there are many regions of DNA that might code for short regions of protein, but these do not look recognizably similar to anything else in other genomes, and they may just be there by chance. However, recent studies have shown that a significant number of such very short 'open reading frames' are indeed transcribed into messages and translated into short proteins (peptides), of unknown function.

There are also regions that clearly do not code for protein, but are copied – transcribed is the technical term – into RNA that may itself have a function. Hundreds of micro-RNA encoding genes have been recognized in genomes in recent years.

Finally, there are regions of DNA that are clearly important – not junk, but they may not be transcribed at all. Instead, they may function to regulate other genes. Whether or not such functional sequences deserve to be called genes is a matter of definition.

So the difficulty in estimating gene numbers is partly an experimental problem, and partly a definition problem. However, it need not really concern us here. The take-home message is that the number of genes in the genomes of complex organisms is not hugely larger than the number in simple organisms. In fact, it is remarkably similar.

A conserved toolkit for animal development

Indeed, comparisons of mammalian genomes with some of our most distant relatives within the animals suggest that a surprisingly large fraction of the complexity of the toolkit to build animals was already present in the last common ancestor of all living animals. The genome that really brought this home was that of a humble sea anemone.

Until this animal was sequenced, most comparisons of complex versus simple animal genomes had been made between vertebrates and two well-known lab models, the fruit fly *Drosophila melanogaster* and nematode worm, *Caenorhabditis elegans. Drosophila* and nematodes have by and large the same families of key developmental control genes as mammals, but the number of different representatives in each of these families is generally smaller in the fly and worm, suggesting that new genes have been invented in the lineage leading to the mammals.

One example of such a family is the family of Wnt genes (pronounced 'wint'), which encode signalling molecules that cells use to communicate with one another. The human genome encodes at least twelve distinct subtypes of Wnt protein, used in different circumstances to do subtly different things in the animal. Flies and nematode worms have only a subset of these Wnt gene subfamilies – it is clear that many of the vertebrate families are missing.

At first this was taken to mean that new types of Wnt genes had arisen in vertebrates, probably by gene duplication and sequence divergence, a process that is known to occur in all genomes. Then the genome of one of our most distant animal relatives was sequenced, that of the sea anemone *Nematostella vectensis.* Most of the Wnt genes missing in flies and spiders were recognizable in this sea anemone genome. That made us think again. We had to conclude that the genes missing in simple animals were not new inventions of the vertebrates, but 'old genes' that must already have evolved their distinct sequence characteristics by the time of the last common ancestor of the animals, but which have since been lost in many animal lineages, including those leading to fly and worm.

Gene loss has turned out to be a surprisingly common story. Another family of important developmental control genes are the homeobox genes. These encode proteins that bind to DNA and regulate other genes, turning them on and off in particular tissues or at particular times. The human genome has hundreds of these homeobox genes – some absolutely central to making the embryo, others playing minor roles in development. But it turns out that we humans and other mammals have lost a few of these homeobox genes over the years. Amphioxus, a small eel-like creature that split from our lineage before the origin of the vertebrates, has retained several that we no longer have.

But our loss is as nothing compared to that of the sea squirts. Sea squirts belong to the same large grouping of animals that contains all the vertebrates, as well as amphioxus – a group technically known as the Phylum Chordata. They retain a tadpole-like larva, but as adults they have settled down to a sedentary life of filtering seawater, throwing away their brains and much else besides. They have lost more than twenty families of homeobox gene from their genome, presumably as a result of drastically simplifying their body plans.

At this point I should correct the impression that the story of genomics is downhill all the way from the sea anemone. There are families of genes that have expanded massively in particular lineages, and which are very important to us. For example, the genes that make the vertebrate adaptive immune system work, or the olfactory receptors that allow us to sense the world. However, the evolution of genomes is clearly not simply a story of ever-greater complexity.

The evolution of cell types

We have seen that the likely ancestor for the animals was some sort of colonial protist related to today's choanoflagellates. In this organism, all cells are equivalent.

That is not true of you or me, or of any multicellular animal. We have specialized cell types. All the cells of an individual have the same genome, but the cells themselves are very different from one another. In general, more complex organisms have more different types of cell. How did these cell types evolve? Recent work from Detlev Arendt and his colleagues at the European Molecular Biology Laboratory (EMBL) is using evidence from genes and genomes to address this question.

The starting point for this discussion is to realize that single-celled organisms are not simple cells – they are quite extraordinarily complex cells that do many different things. For example, in a unicellular organism that swims towards light, the same cell senses light, contains shading pigment to make that sensing directional, and executes the swimming function. Our specialized cell types are in some respects much simpler. They do one thing very well – photoreceptors in the eye sense light, but separate lenses focus the light; nerve cells conduct impulses, muscle cells contract to allow motion. Many different cells are needed to accomplish even a simple response to a stimulus.

Arendt studies a very simple multicellular animal – a marine worm called *Platynereis dumerilii*, a close relative of the ragworms that fishermen dig out of muddy beaches to use as bait. This worm has a swimming larva of a few hundred cells, which has a single row of swimming cells around the middle, connected by a single nerve to a pair of eyes, each of which is made up of just two cells, a photoreceptor that senses the light and a shading pigment cell that makes the photoreceptor directionally sensitive. Even this is enough to show the first signs of complex behaviour. The photoreceptor cells change the beat of the ciliated swimming cells such that, depending on where the light is, the animal will spiral towards it. A similar pattern of behaviour is seen in many small marine larvae. Arendt has proposed that the diversity of cells in this swimming system arose by a division of labour from an initially much more complex cell. Not the way we normally think of evolution at all.

He points out that some very simple animals still have multifunctional photosensitive swimming cells in which all the aspects of the light response are combined, much as they are in unicellular organisms. The eyes of some box jellyfish provide an example. Arendt suggests that the first step towards building a nervous system involved partitioning these functions into distinct cell types, with each ancestral cell type giving rise to multiple descendant cell types that executed only a subset of the ancestral functions. Here again, genes provide supporting evidence. The set of genes expressed in each cell type provides a 'molecular fingerprint' that we might expect to reflect function, but which may also reflect ancestry. Arendt finds that the 'molecular fingerprints' of the two different cells in the larval eyes are more similar to one another than either is to the cell with equivalent function in the adult eye. He suggests that this is evidence that these are sister cell types, related by evolutionary descent.

Remarkably, by collecting such molecular fingerprints in a range of species, Arendt and his colleagues are beginning to be able to draw comparisons between the cells of this simple worm and the cells in own brains, and to suggest that both the vertebrate nervous system and the worm nervous system are built around a set of rather conserved units, where we can still read, in the way the genes are expressed, the history of how a few ancestral cell types have changed and diversified. This is still quite speculative, but I think that it represents a most interesting new way of using genetic information to understand how animal complexity evolved.

The organization of cells in space

It is not enough to generate cell types. To make an animal, the right cells must be put in the right places. This is the invention of embryogenesis.

Some animal species are quite extraordinarily good at putting cells precisely in the right place. It was realized in the 1890s that snail embryos repeat from individual to individual the same exact pattern of cells. Each cell could be given a specific name and number, which would predict what it would give rise to in the later animal.

We vertebrates are nothing like as precise as snails when it comes to putting cells in the right place to make our own embryos, but all animals have to pattern their embryos precisely enough, so that all cells know where they ought to be, whether they are part of the head or part of the tail. One of the surprising insights that came from early comparisons of how genes are used to build embryos was that radically different organisms appear to do this in remarkably similar ways. This is an old story, but I will illustrate it with some relatively new work of Chris Lowe and his colleagues, comparing vertebrates with acorn worms. Acorn worms are not the most familiar of marine worms, but they are of particular interest because of all the worm-like creatures, they are the most closely related to us; although not chordates, they are in the same great branch of the animal kingdom that includes the vertebrates and other chordates.

Acorn worms look nothing like vertebrates. Indeed, you might be hard pressed to say how you would compare the embryo of an acorn worm with that of a developing fish or mouse (Figure 3.5). But when you look at the critical set of genes that control the specification of the early embryo, the genes known as transcription factors that make particular sorts of cells by turning other genes on and off, it turns out that there is a quite extraordinary degree of similarity in the early patterning of an acorn worm and a typical vertebrate. (All vertebrates are rather similar at this early stage in development.)

The bars in the figure indicate the domains of the embryo where each particular gene is expressed. The head end of a vertebrate expresses one subset of genes, and those same genes are expressed in the proboscis of the acorn worm; then there is the mid-brain region of the vertebrate, which expresses a different set of genes in the early embryo. Those genes are expressed in the collar of the acorn worm. Finally in the rest of the vertebrate body there are the famous Hox genes, which are expressed in sequence from the back of the head to the tail. They tell the neck vertebrae to be different from the chest vertebrae to be different from the lumbar vertebrae. The same set of genes is expressed in a similar way to do something – we do not know what – in the very boring back end of the acorn worm.

This conserved head-to-tail co-ordinate system is present, not just in these two representatives of one lineage, but largely conserved across the immense diversity of animal forms. I could have drawn similar pictures

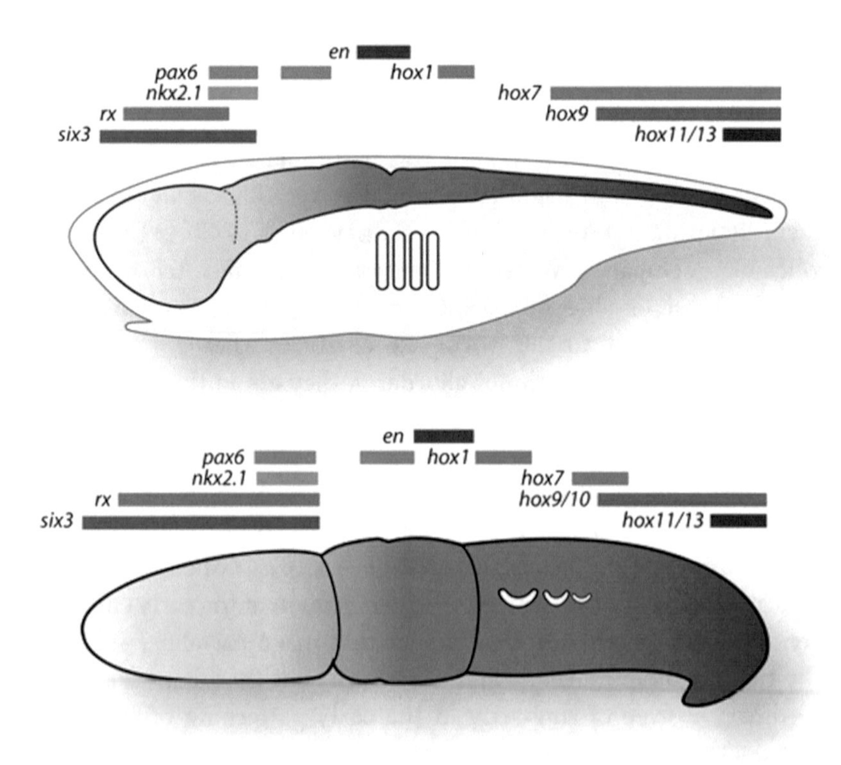

FIGURE 3.5 This schematic diagram compares a vertebrate embryo (above, head to left) with an acorn worm (below). The acorn worm has no obvious head, but is divided into three regions, a proboscis (left), a collar and a so-called tail. Although there is little morphological similarity between these animals, the same set of genes is expressed in specific domains along the anterior to posterior (head-to-tail) axis of the body, giving each region its particular identity. (Picture courtesy of Ari Pani and Chris Lowe.)

with the ragworm, or with a fruit fly. And analogous comparisons suggest that something similar is true for our back-to-belly axis, too, but with an intriguing twist, and a fascinating historical context. The story begins with Geoffroy Saint-Hilaire, a famous French biologist of the early nineteenth century, who proposed a relationship between the organization of the vertebrate body plan and the organization of the 'lower animals', as he would probably have called them.

Vertebrates have a dorsal nerve cord and a ventral aorta. If you turn this structure upside down you get an animal with a ventral nerve cord and a dorsal aorta, a pattern seen in flies and ragworms. Saint-Hilaire proposed that vertebrates and many invertebrates share a common ancestral organization, but that there had been an inversion of the back-to-belly axis in vertebrates. Baron Georges Cuvier, the most powerful zoologist of his time, rubbished Geoffroy Saint-Hilaire's argument, and the idea died, though it was resurrected repeatedly over the next 150 years.

Now, exactly the same sort of data that I discussed for the head-to-tail axis, suggests that Cuvier was wrong. Vertebrates and almost all other animals use the same set of genes to pattern their back-to-belly axis, but vertebrates have indeed reversed the orientation. The same family of molecules specifies the middle of the belly of a fruit fly, where their nerve cord forms, and the middle of our back, where our nerve cord develops. This idea does involve the slight complication that our mouths cannot be in the same position that they were when we first evolved. There are endless arguments about that. But Geoffroy Saint-Hilaire was probably right.

The genomic basis for animal diversity

If so much of the genetic make-up of all animals is conserved, what is generating the wonderful diversity of animals that we see today? We have to change the way we think about genes to approach this. The idea that genes make proteins is the biochemist's view of a gene. It focuses on what happens downstream, when the gene is transcribed to message, and the message translated.

But there is something that happens upstream, something that happens with the bits of DNA that are not actually making the protein, but are telling the system where and when to make that protein. There are regulatory elements upstream of genes that receive signals. They receive signals about the state of their own cell, and from neighbouring cells. They may also receive signals from hormones, from light, from many different sources, and they integrate this information to produce an output – whether that gene is to be on or off.

From this perspective, a gene is as much an information processor as it is part of a machine tool. Eric Davidson, an influential biologist at the California Institute of Technology, has championed this idea. Think of

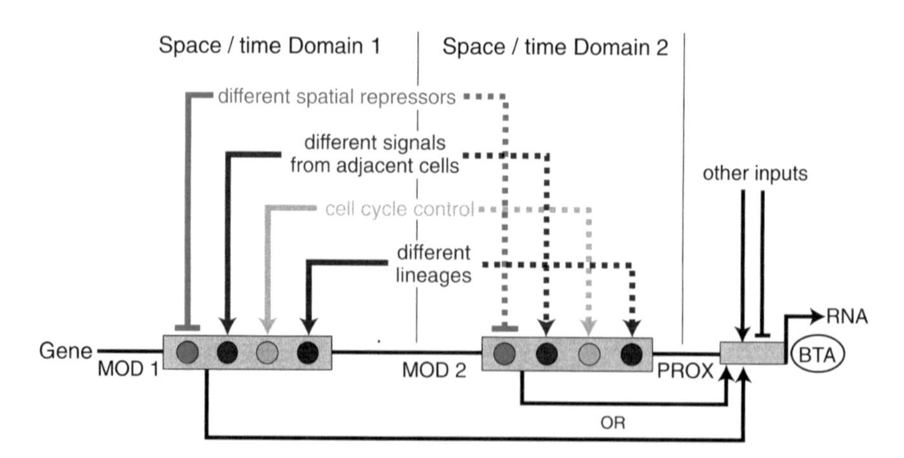

FIGURE 3.6 A gene is not only the DNA that encodes for a protein or RNA product; it also includes the regulatory sequences that specify where, when and how much of that product should be made. Eric Davidson has pioneered the view that genes should be seen as information processors, integrating many different sorts of information. The logical wiring of the genome specifies how all of these different processors interact in the gene networks that control development. (Figure reproduced from E. H. Davidson, *Genomic Regulatory Systems*. San Diego: Academic Press, 2001.)

genes as microprocessors, he argues (Figure 3.6). The way genes talk to one another is what matters. They interact as a network, and the genome implies a circuit in the sense that the activity of one gene will affect another component in the system, will turn it on, will turn it off, will tell a cell that it should be making hair now and then stop making hair or make it a different colour in a little while. It is these essential but complex interactions between genes that are critical for generating diversity. The genes may be conserved, but the connections between them may be much more variable in different animals.

The question of how interactions between genes change to generate animal diversity is the focus of a network of labs called Evonet, to which my lab belongs. We study many of the animals I have talked about and more, seeking to understand how to read in genomes the interactions between genes that control development.

The problem is that we do not know the code yet. We cannot simply take a computer and ask it to tell us which genes one particular protein will control. There are techniques that are beginning to figure that out.

Eileen Furlong's lab, for example, is finding experimentally all the sites in a genome where the controlling protein made by one particular gene is able to bind and regulate other genes. But computers cannot do that at all well yet. It is still a real challenge for genome biologists to understand the logic that is implicit in genomes.

My own particular interest is animal segmentation – building bodies from repeated structural units. That is why I am interested in centipedes, which make an unusually large number of segments (Figure 3.7). Segmentation is something that is seen in several of the most successful animal groups. I am interested in knowing whether segmentation evolved once at the beginning of animal history, or independently several times; whether similar molecular machinery is used in different cases, and how, when you have an ancestral machinery that was good at making segments rather slowly, as in the centipede, it evolved to make a fruit fly that grows from embryo to adult before a rotting banana has time to dry up, and can make all of its segments in just a few hours.

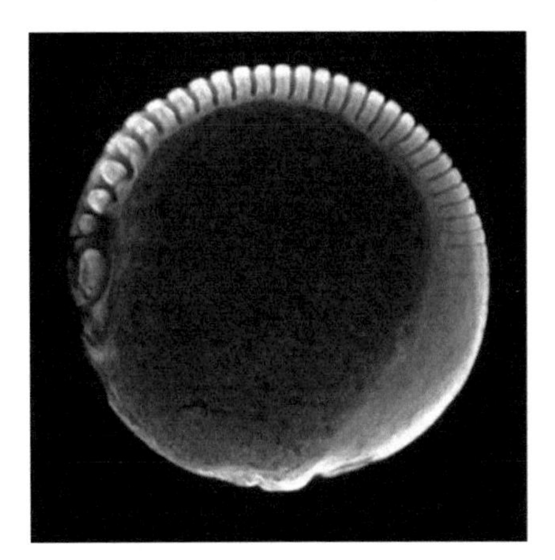
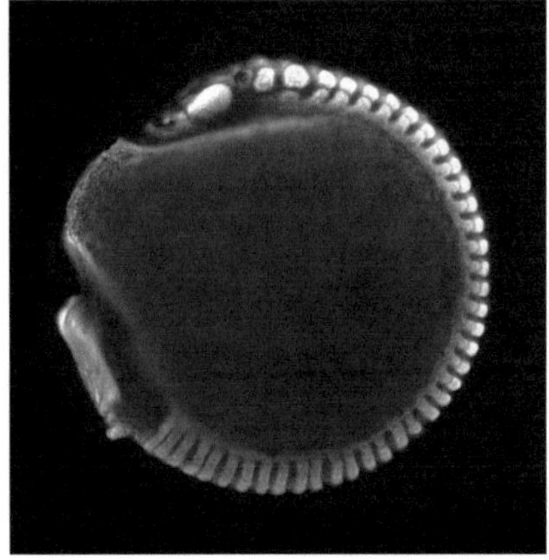

FIGURE 3.7 As the embryo matures, cycles of repeated gene activity sweep through an unpatterned sheet of cells, visible at the lower right of the younger embryo on the left, generating the molecular patterns that control cell behaviour, and hence the subdivision of the body into segments, each of which will bear a pair of legs with their associated muscles and nerves. In the older embryo on the right, most of the segments are demarcated by grooves in the body wall, but legs have yet to form. (Picture courtesy of Carlo Brena.)

I cannot here discuss all that we know about the segmentation machinery in flies. I have spent twenty years of my career looking at some of those gene networks. We know a great deal, though still not enough to tell a computer how to build a fly. What I would love now is to understand how subtle changes to those gene networks build not a fly, but a shrimp, or a spider, or that most beautiful embryo of a centipede.

I was told many years ago that developmental biologists start out by looking at their animal and saying 'I'm going to understand this.' As they come closer to the end of their career they look at their embryos and simply say, 'Aren't they wonderful!' Maybe I am getting close to that stage of my career now, but I still think that there is much we can to understand how genomes make animals.

Further reading

For a current overview of animal relationships in the light of genome information, see Peter Holland's *The Animal Kingdom: A Very Short Introduction* (Oxford: Oxford University Press, 2011). For an accessible introduction to ideas about how genomes lead to the diversity of life, see Sean Carroll's *Endless Forms Most Beautiful: The New Science of EvoDevo* (London: Weidenfeld and Nicolson, 2005; paperback 2006).

4 Artificial life

CHRIS BISHOP*

Living organisms are extraordinary. They have capabilities which far exceed any present-day technology, and it is therefore inevitable that scientists and engineers should seek to emulate at least some of those capabilities in artificial systems. Such an endeavour not only offers the possibility of practical applications, but it also sheds light on the nature of biological systems.

The notion of artificial life can take many diverse forms, and in this article we will focus on three aspects: modelling the development of structure in living systems, the quest to create artificial intelligence, and the emerging field of synthetic biology. All three topics reveal surprising, and sometimes remarkably deep, connections between the apparently disparate disciplines of biology and computer science. There is something else which links these three strands: the Cambridge mathematician Alan Turing (see Figure 4.1) whose birth centennial we celebrated in 2012.

It is widely acknowledged that Turing laid many of the foundations for the field of computer science, although amongst the general public he is perhaps best known for his role in breaking the Enigma and other cyphers at Bletchley Park during the Second World War (Hodges 1992). What is perhaps less widely appreciated is that Turing also made important contributions to biology. As we shall see, each of the three topics discussed in this paper builds on a different seminal contribution made by Turing.

* I am very grateful to many colleagues for their assistance in preparing this lecture, including Luca Cardelli, Andrew Phillips and Matthew Smith.

FIGURE 4.1 Alan Turing (1912–1954).

Morphogenesis

One of the most intriguing challenges in biology is to understand the mechanisms by which an organism acquires structure and form during its development. This is known as 'morphogenesis' (the 'creation of shape'). A visually striking, and very familiar, example of structure in a living organism is given by the spots on a leopard (Figure 4.2).

The fact that a leopard has spots rather than, say, stripes, is determined genetically. However, the precise size, shape and location of individual spots do not seem to be genetically controlled but, instead, emerge through a mechanism for creating 'spottiness'. One of the earliest attempts to elucidate such a mechanism was given by Alan Turing. In 1952, just two years before his untimely death, he proposed a chemical basis for morphogenesis, together with a corresponding mathematical analysis

FIGURE 4.2 Leopard with spots. © Shutterstock.

(Turing, 1952). His idea was that spatial structures, such as the spots on the leopard, arise from the interaction between specific chemical and physical processes in an example of spontaneous pattern formation, also known as 'self-organization'.

There are many examples of self-organizing systems. For example, when a magnetic field is applied to a ferrofluid, a regular array of spikes is formed, as illustrated in Figure 4.3.

The ferrofluid consists of a colloidal suspension of nanometre-scale ferromagnetic particles dispersed in a liquid. In the absence of a magnetic field, the ferrofluid behaves like a regular, slightly viscous liquid. When it is placed into a magnetic field, however, the ferrofluid organizes itself into regular spikes, the size and spacing of which depend on the strength of the magnetic field. This period structure arises from the interaction of the ferrofluid with the magnetic field, as neither the fluid nor the magnetic field individually has any intrinsic structure of this kind. The external field magnetizes the fluid, which then forms into spikes as this leads to a lowering of the magnetic energy. This phenomenon of energy

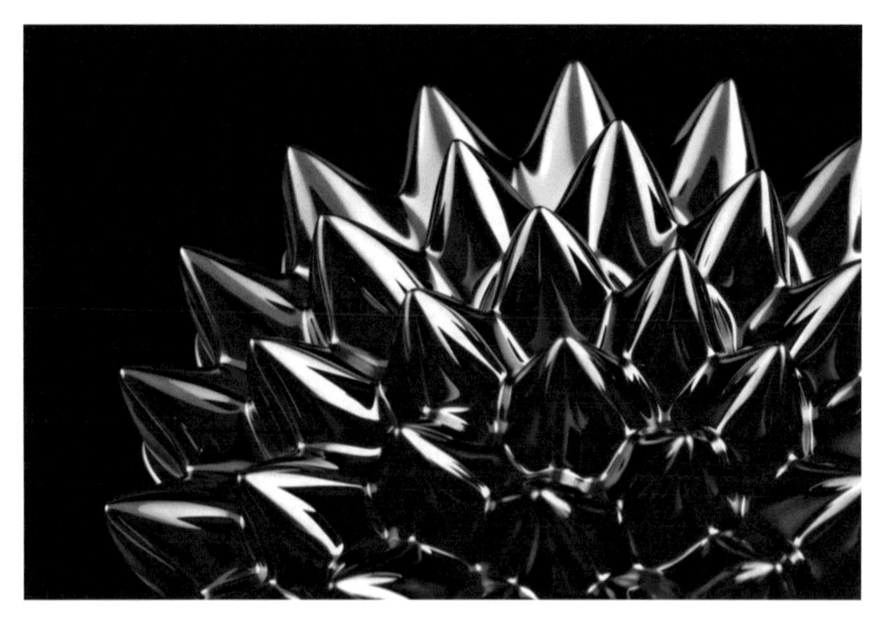

FIGURE 4.3 Spikes in a ferrofluid resulting from the application of a magnetic field. © Shutterstock.

minimization is familiar from the behaviour of two bar magnets, which, if allowed to move freely, will align with their poles pointing in opposite directions and will then come together. The ferrofluid likewise adopts a configuration that minimizes the overall energy of the system, with the spikes growing until the reduction in magnetic energy is balanced by the increase in gravitational and surface tension energies. As long as the magnetic field is sufficiently strong, the formation of spikes is energetically favourable.

Clearly a different process must be responsible for pattern formation in biological systems. The specific mechanism proposed by Turing is known as a 'reaction-diffusion system'. Here 'reaction' refers to chemical reactions, while the concept of diffusion can be explained using the illustration in Figure 4.4.

This diagram depicts a molecular-level view of a situation in which there is a spatial gradient in the concentration of some substance. In this case the concentration gradient is from left to right. The molecules

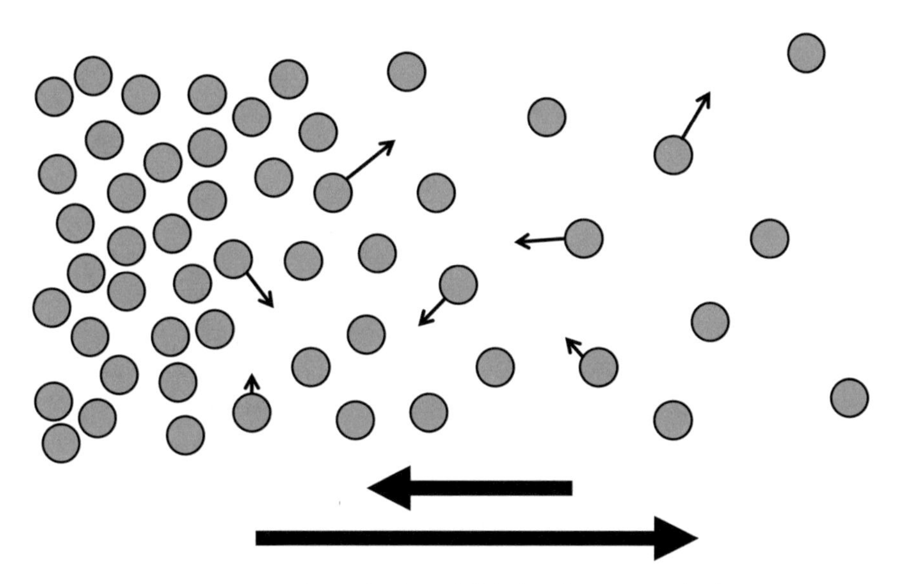

FIGURE 4.4 The mechanism of diffusion. See text for details.

themselves are in constant, random motion as a result of being at a non-zero temperature (heat is just the random motion of molecules) and this is illustrated by the arrows attached to some of the molecules. Each molecule is equally likely to move to the left as to the right. However, there are more molecules on the left of the picture than on the right, so on average there will be a greater flow of molecules from left to right compared to the flow from right to left. Overall, therefore, there is a net flow of molecules from left to right, and over time this will act to reduce the concentration gradient. This effect is easily demonstrated experimentally by carefully adding a few drops of food colouring to a beaker of water, without stirring. Over the subsequent hours and days the colouring will gradually spread through the water until the colour of the solution is uniform. As the gradient approaches zero so too will the net flow of molecules. This statistical property whereby on average there is a flow of a substance in the opposite direction to a concentration gradient, as a result of the random thermal motion of the molecules, is called diffusion.

Now consider a situation in which there are two or more species of molecule, each having gradients in concentration, in which the species can undergo chemical reactions with each other. Because chemical reactions transform molecules of one type into molecules of other types they effectively change the concentration and hence the concentration gradients. Conversely, the rate of chemical reaction depends on the concentrations of the reacting species. The result is a complex interaction between reaction and diffusion.

Turing analysed a particular class of reaction-diffusion systems mathematically and discovered that, under appropriate conditions, such systems can exhibit a wide variety of behaviours, but that they always converge to one of six stable states (Kondo et al. 2010). The first of these involves concentrations that are uniform in space and constant in time, and are therefore relatively uninteresting. The second consists of solutions which are uniform in space but which oscillate through time, as seen in circadian rhythms and the contraction of heart muscle cells. The third and fourth classes of solutions consist of 'salt and pepper' patterns of high spatial frequency which are either constant (third class) or which oscillate through time (fourth class). Of these, the former are seen in neuro-progenitor cells in the epithelium of *Drosophila* embryos, while no examples of the latter have yet been identified in living organisms. The fifth class exhibits travelling waves having spatial structure which evolves through time. Beautiful demonstrations of such solutions can be performed in the laboratory using simple chemistry, while biological examples include the spiral patterns formed by clusters of the amoeba *Dictyostelium discoideum*. The sixth class of solutions consists of spatial structures that are constant in time, and have become known as 'Turing patterns'. These are remarkable solutions since the patterns are stable, and can even regenerate following an external disturbance. They represent a dynamic equilibrium in which the effects of reaction and diffusion are balanced, leading to a stationary structure. Such structures are self-organizing and do not require any pre-existing spatial information. There is now good evidence to support the role of this mechanism in creating a variety of biological patterns including complex patterns of seashells, the patterning of bird feathers, and the impressive diversity of vertebrate skin patterns.

We do not have to have a literal reaction-diffusion system for Turing's mechanism to operate. More recently it has become clear that the conditions for Turing pattern formation are more general, and require only

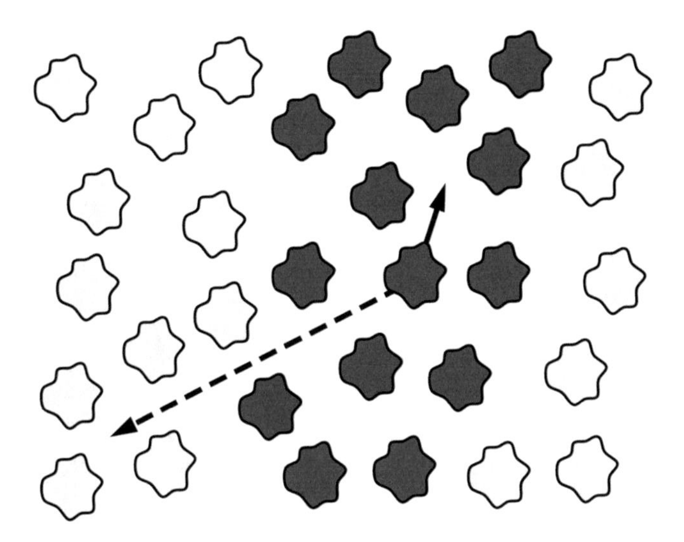

FIGURE 4.5 General mechanism for the formation of Turing patterns.

the presence of two competing effects having rather general properties (Kondo et al. 2010). One effect acts over a short range and encourages regions to be the same (in terms of concentration, colour or some other characteristic) while the other effect acts over a longer range and encourages regions to be different. This can most easily be illustrated using the simple example shown in Figure 4.5, which considers the specific example of cells which can communicate via chemical signalling.

We shall suppose that each cell can take one of two colours, dark or light. Each cell can also send two chemical signals which can be detected by other cells. One of these signals (indicated for a particular cell by the solid arrow in Figure 4.5) acts over a short range and encourages other cells to take on the same colour as the transmitting cell. The second signal (indicated by the dashed arrow in Figure 4.5) acts over a longer range and has the converse effect, encouraging other cells to have the opposite colour to the transmitting cell. Initially, the cells have randomly assigned colours, but the effect of the interactions due to the chemical signalling is to cause the ensemble of cells to reach a stable configuration. If the parameters of the signalling are suitable, then this stable configuration will exhibit a Turing pattern, such as the strips indicated in the figure.

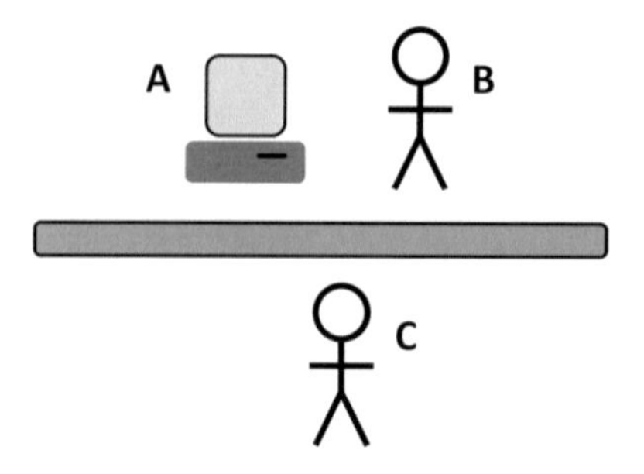

FIGURE 4.6 Schematic illustration of the 'Turing test'.

We shall return to a discussion of Turing patterns later, but for the moment we consider another of the great intellectual challenges tackled by Turing.

Artificial intelligence

Visions of intelligent machines go back to antiquity, but it is with the development of programmable digital computers that the realistic possibility arose of building such machines. Alan Turing's seminal work on the theory of computation laid the foundations for modern digital computers, and Turing himself was also fascinated by the challenge of building intelligent machines. One obvious problem is that there is no simple definition of what it means to be intelligent, and therefore no simple criterion for success. Turing addressed this by introducing an operational procedure for determining whether a machine is intelligent (Turing 1950). This has become known as the 'Turing test', and is illustrated in Figure 4.6.

The essential idea is to compare the artificial system with a human on a series of tests, and if it proves impossible to distinguish which is which, then the machine is deemed to be intelligent. This is achieved by separating the interrogator C from the computer A and the comparison human B (Figure 4.6). The interrogator can communicate with A and B only through a text-based system such as a keyboard and screen, and initially does not know which of A and B is the computer and which is the human. This use of text as a medium for communication was chosen to bypass issues with the computer having to interpret and synthesize audio speech. The interrogator can ask questions of both A and B and, based on the answers, must decide which is human and which is the machine. If the interrogator is unable to distinguish them then, according to the Turing test, the machine is deemed to be intelligent.

The Turing test has generated intense discussion and controversy over the years. While it has the benefit of being an operational definition, it also has some clear limitations. In particular, there are many other forms of 'intelligence' exhibited by living systems which are not captured by the test. For example, a chimpanzee could never pass the Turing test because it cannot read and write text, but it has numerous capabilities (recognizing objects, locomotion, fine motor control, planning, use of tools, and many others) that it could be very beneficial to emulate in machines. Moreover, from a technological perspective, imitation of humans is, of itself, only of limited interest. If the interrogator asked a computer to solve a complex arithmetical problem and the answer came back correctly a split second later it would be clear that this was a machine not a human, and so the computer would have failed the Turing test, even though it was outperforming the human! It has been said that 'aeroplanes are measured by how well they fly, not by how accurately they mimic birds', and so it is with machines.

Artificial intelligence became a popular field of research from the mid-1950s, with much of the effort focused on techniques based on hand-crafted solutions, often involving logical processes which manipulate symbolic representations of the world. A specific approach, which became popular in the 1970s, was based on rules derived from a human expert through a process called 'knowledge elicitation'. The resulting artificial instantiation of those rules was known as an 'expert system'. To give a concrete example, suppose we wish to build a system that can diagnose a human disease, given a list of the patient's symptoms. The computer might be fed with hand-crafted rules of the form 'if the patient has a high temperature and low blood pressure then . . . ' in which the rules are obtained from human experts. For a specific application domain, a set of rules would be compiled, known as a 'knowledge base', and then a piece of software called an 'inference engine' would invoke appropriate rules in answering a specific query. Although this approach of hand-crafting intelligent behaviour has had some worthwhile successes in niche applications, it has broadly failed to deliver a general framework for creating machine intelligence. The fundamental problem seems to be that, except in very specific domains, the real world cannot effectively be captured by a compact set of human-expressible rules. Attempts to capture exceptions to rules using additional rules, leads to a combinatoric explosion of exceptions-to-exceptions.

Since the mid-1980s the most prevalent, and most successful, approach to creating intelligence in machines is based on a very different paradigm: 'machine learning' (Bishop 2006). Instead of programming the computer directly to exhibit a particular form of intelligent behaviour, the computer is programmed to be *adaptive* so that its performance can improve as a result of experience.

In the case of a medical diagnosis system, the 'experience' might consist of a database of examples, each of which comprises a list of observed symptoms along with a diagnosis of the disease provided by a human expert or from some other form of ground truth such as a blood test. Such a database is known as a 'training set'. A simple form of machine-learning technique in this case could consist of a non-linear mathematical function that maps symptoms as inputs to diseases as output. This function is governed by some adjustable parameters, and the computer has an algorithm for optimizing these parameters so as to make the accurate prediction of the corresponding disease for each set of symptoms. One well-known class of such non-linear functions, that was very prevalent in the 1990s, is called a 'neural network'. Today there are hundreds, if not thousands, of different machine-learning techniques.

It is worth noting that such systems can easily be made to achieve good predictions on the training data examples, but that the real goal of the system is to be able to make good predictions for new examples that are not contained in the training set. This capability is called 'generalization'. Essentially it requires that there be some underlying regularity in the data, so that the system can learn the regularity rather than the specifics of each instance, and hence can generalize effectively.

Techniques of this kind have achieved widespread success in many different application domains. A recent example is the Kinect full-body tracking system used on the Xbox games console, which was launched in November 2010 and which set a new Guinness world record as the

FIGURE 4.7 The thirty-one body parts to be recognized by Kinect. © Kinect.

fastest-selling consumer electronics device of all time. This is the first piece of consumer electronics to provide real-time tracking of the entire human body, and has been deployed initially as a hands-free games controller. It is based on a special type of infra-red camera that measures depth rather than intensity, so that each pixel in each frame of the captured video represents the distance from the sensor to the corresponding point in the scene. Unlike a conventional camera, this sensor thereby has a three-dimensional view of the world, which eases the task of tracking the human body in 3D space.

The fundamental problem that needs to be solved in order to use the sensor to track the human body is the ability to recognize the various parts of the body in real time using the depth data. This was solved using a machine-learning approach as follows. First, thirty-one regions of the body are defined, as illustrated in Figure 4.7.

Then in each frame of the video, the games console must take the depth image and, for each pixel in the image, classify that pixel according to which of the thirty-one body regions it corresponds to (or whether it belongs to the background). This is solved using a machine-learning technique called 'random forests of decision trees' (Shotton et al. 2011). In order to train the machine-learning system, a training set was created consisting of 1 million body poses, each with a 3D depth image and each having every pixel labelled with the correct body region. In order to ensure that the training set contains a representative range of body positions, the poses were collected using a Hollywood-style motion capture suite. An actor wears a special suit which has high-visibility markers attached at key points, and these are viewed by multiple conventional cameras placed in different positions looking at the actor. By correlating the data from the cameras the 3D body pose can be inferred. These body poses are then used to generate a synthetic training set which accounts for variations in body size and shape, and clothing, as well as artefacts introduced by the depth camera itself.

The million examples in the training set are then used to tune the parameters of the random forests of decision trees. This is done using a large array of PCs, and is a computationally very intensive process. Once the system has been successfully trained, however, it can be deployed on the games console where it can classify new pixels very quickly. In fact it can classify all of the relevant pixels in each frame of the video (at some thirty frames per second) in real time while using less than 10 per cent of the processing power of the Xbox console. Although Kinect was originally conceived for use as a games controller, it is being tested in a variety of user-interface applications, for example in operating theatres to allow surgeons to manipulate digital images of the patient without contact with physical mice or keyboards that would compromise sterility.

Another impressive example of the power of machine learning took the form of the Grand Challenge organized by the US Defense Advanced Research Projects Agency (DARPA). The goal was for a fully autonomous vehicle to drive a distance of 240 km along a pre-determined path through the Mojave Desert, with a prize of US$1 million. The competition first took place in March 2004, and none of the vehicles managed to complete the course, with the most successful vehicle managing to cover less than 12 km; the prize was not awarded. The following year, the competition was repeated, this time with a prize of US$2 million and an even more challenging course. Five vehicles managed to complete the course, with the winning team from Stanford receiving the prize. Their algorithm used machine learning to solve the key problem of obstacle avoidance, and was based on learning from a log of human responses and decisions under similar driving conditions.

Following the success of the 2005 race, DARPA held the Urban Challenge in 2007, involving a 96 km urban route, to be completed in under six hours, with a total prize of US$3.5 million. Vehicles had to obey all traffic signals and road markings while also dealing with the other robotic vehicles on the course. The winning entry was from Carnegie Mellon University, completing the course in just over four hours. Autonomous vehicle technology for cars to replace human drivers offers the promise of reduced numbers of accidents and fatalities, increased efficiency, and the opportunity for drivers to make better use of their time (e.g. to work, sleep or watch a film) while the car drives itself.

DARPA recently announced a new Robotics Challenge which will involve the solution of complex disaster-relief tasks by humanoid robots in situations considered too dangerous for humans to operate. Unlike previous challenges, the teams will be provided with identical humanoid robots, and so the focus will be on the artificial intelligence software for controlling them. The robots will be expected to make use of standard tools and equipment commonly available in human environments, as illustrated conceptually in the DARPA image shown in Figure 4.8.

As our final example of the rapid progress of artificial intelligence through the use of machine learning, we look at the widely publicized participation in 2011 of the IBM computer system 'Watson' in the US television quiz programme *Jeopardy!*

The *Jeopardy!* show was created in 1964, and over 6000 episodes have been broadcast. Its format is slightly unusual in that the game host poses trivia puzzles in the form of answers to which the contestants must provide the appropriate questions. A special series of three programmes was run to host this contest in which Watson played against the two leading human players from previous years. In preparation to play *Jeopardy!*, Watson had been fed with 200 million pages of information from encyclopaedias (including the whole of Wikipedia), dictionaries, thesauri, news articles and other sources, consuming a total of four terabytes of disk storage.

FIGURE 4.8 Conceptual image of the DARPA humanoid Robotics Challenge, depicting two robots tackling an accident in a chemical plant. © DARPA.

However, Watson was not connected to the internet during the game. Watson's software is based on a complex system of interacting modules that process natural language, retrieve relevant information, and represent and reason about knowledge, with machine learning playing a key role. The Watson hardware was capable of processing around 80 trillion computer instructions per second, and was represented on stage by an avatar placed between the two human contestants, the actual hardware being far too large to fit on stage.

During the game, Watson's three highest-ranked potential responses were displayed on the television screen, but Watson only buzzed when the confidence of the highest-ranked answer exceeded a suitable threshold. Watson then provided a response using a synthesized voice. In a combined-point match spread over three episodes, Watson was the comfortable winner and collected the US$1 million first prize which was subsequently donated to charity. This head-to-head competition with humans in a natural language setting is somewhat reminiscent of the original Turing test. It represents a remarkable development in artificial intelligence because natural language has long been seen as a highly challenging domain for computers. Today machine learning has become one of the most active and important frontiers of computer science.

Synthetic biology

Our third and final viewpoint on artificial life reveals some deep connections between biology and computer science, and again relies on some key insights from Alan Turing. Research in molecular biology has revealed extraordinarily complex networks of biochemical interactions in living cells. As the details of such networks have been identified, it has become increasingly clear that much of the functionality of these networks is concerned with information processing. It is not just computer scientists who hold this view. In his recent talk at the Royal Institution, Sir Paul Nurse, Nobel Prize winner and President of the Royal Society, spoke of the 'Great Ideas of Biology', which he listed as the cell, the gene, evolution by natural selection, and biology viewed as complex chemistry. He then suggested that the next great idea of biology would be the recognition that life is 'a system which manages information'.

The view of molecular biology as information processing is much more than simply a convenient analogy, and to understand the depth of this viewpoint we need to look at a few key concepts in molecular biology and also to turn again to the work of Turing. At the heart of the biomolecular machinery of the living cell is DNA, or deoxyribonucleic acid. This is a molecule with a double helix structure in which each strand of the helix comprises a sequence of bases, each of which is either C (cytosine), A (adenine), G (guanine) or T (thymine). If we imagine unwinding the helices we would obtain a structure of the form shown conceptually in Figure 4.9.

We can see immediately that DNA is a digital information storage system in which the information is represented by the sequence of bases. Note that the two strands of DNA are complementary, in that a G always occurs opposite a C, and a T always occurs opposite an A. Thus the two strands carry the same information, but in complementary

G C T G A A G T C A

C G A C T T C A G T

FIGURE 4.9 The two strands of DNA showing the complementary sequences of bases.

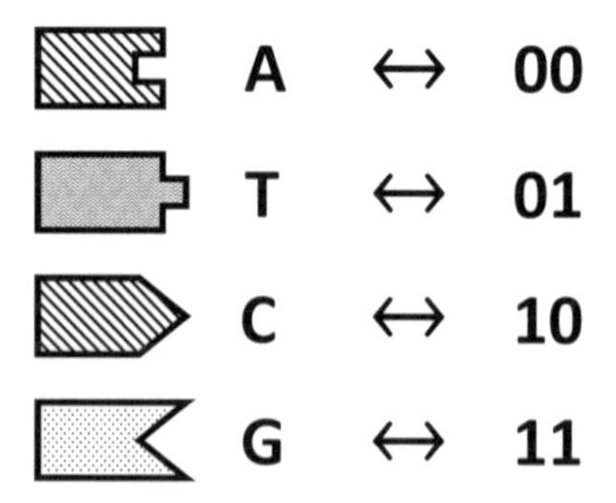

FIGURE 4.10 An association between bases and pairs of binary digits.

representations. The same information could equally well be represented in the binary representation used by computers, for example by making the (arbitrarily chosen) associations shown in Figure 4.10.

Thus the sequence of bases GCTGAC would be represented as 111001110010. Only ten base pairs were shown in Figure 4.9. The DNA for a human has around 3 billion pairs of bases, and this DNA is contained in almost all of the 100 trillion cells in the human body. The DNA in each cell therefore stores around 6 billion bits of information, or roughly 750 megabytes, which is roughly equivalent to data stored on an audio CD.

The DNA contains genetic information that is inherited by an organism's offspring and which controls many aspects of the development and function of the organism. Specific sections of the DNA, called genes, control the production of proteins through a two-stage mechanism in which the sequence information is first copied to a related type of molecule called 'messenger RNA' in a process called 'transcription'. The messenger RNA is then used to synthesize proteins by structures called ribosomes in a process called 'translation'. Although essentially every cell in an organism contains the same DNA, in a particular cell and at a particular time, only some of the genes are *expressed*, with others essentially dormant. The degree of expression of a particular gene is controlled by an associated section of the DNA called a 'regulatory region'. Proteins binding with the regulatory region can alter the degree of gene expression. Here we see the beginnings of a very complex feedback system in which DNA controls the production of proteins which themselves can influence the production of other proteins. We therefore see that the DNA is much more than simply a static information store. It is a key component in a highly complex dynamical system with multiple feedback loops in a process that strongly resembles computation in digital computers.

By modifying the DNA the properties of the organism can be changed. This has been done for thousands of years through selective breeding in order to produce genetically altered organisms having desired characteristics (e.g. a stronger horse or a more nutritious variant of wheat). With the discovery of the molecular basis of inheritance, came the opportunity to do this more specifically and more rapidly, leading to the field of *genetic engineering*. This involves copying a gene from one organism and inserting it into the DNA of another organism, for example, a gene from a bacterium which codes for an insecticidal protein can be inserted into the peanut plant, thereby protecting its leaves from attack by the larvae of the cornstalk borer. When the larvae eat the leaves they are killed by the insecticide which is now being synthesized by the peanut plant.

Since then, significant advances have been made in our ability both to sequence and to synthesize DNA. We are now no longer restricted to cutting and pasting single genes from one organism to another, but can read and write whole genomes. The power of this approach was first demonstrated by Craig Venter and colleagues, who in 2010 took the DNA from a species of bacteria called *Mycoplasma mycoides* and ran it through a machine that measured the sequence of its 1.2 million base pairs, reproducing the DNA sequence information in digital form on a computer. Next they inserted various watermarks including quotations from famous people, and an email address. The resulting sequence was then sent to another machine, somewhat analogous to an inkjet printer, but using the four DNA bases rather than four colours, which synthesized the corresponding DNA. They then took a different species of bacteria called *Mycoplasma capricolum* and removed its DNA and replaced it with the synthetic DNA. The resulting organism, nicknamed 'Synthia', multiplied and exhibited all the characteristics of the original *Mycoplasma mycoides.* Overall this project took twenty scientists around ten years, at a cost of US$40 million.

While this was a remarkable technical achievement, few would regard Synthia as truly an artificial life-form, as it relied both on an existing host cell with its bimolecular machinery, and on the known DNA sequence of the original organism. It is somewhat analogous to deleting the operating system on a computer and replacing it with the operating system from a different brand of computer, which can be done with very little understanding of how the operating system actually works. A much more exciting, and substantially more challenging, goal is to produce modified organisms by redesigning parts of the DNA sequence rather than simply by copying genes from one organism to another. To see how we might achieve such a goal, it is useful to look more closely at the nature of information processing and computation.

We are familiar with a wide variety of computational devices, from mobile phones to desktop PCs, and from pocket calculators to supercomputers. Clearly these machines differ in speed, physical size, cost and other attributes, but do they also differ in the kinds of computational problems which they can solve? The answer to this question was given by Alan Turing and is rather surprising. Turing first showed that there are computational problems which no computer will ever be able to solve. As an example he described the 'halting problem' in which a machine is presented with an example of a computer program and must decide if that program will halt after a finite number of steps or if it will run for ever. For many programs this is easy to solve, but Turing showed that no machine can exist which can provide the correct answer for all possible programs. Thus, there is a limit to the computational capability of computers! Turing then went on to show that there can exist a computer which can reach this limit, that is, one which can solve any problem which is computationally soluble. This is known as 'Turing universality', and it follows that the set of problems which can be solved by one universal computer is exactly the same as the set which can be solved by any other universal computer. There is a technical caveat that the computers each have access to unlimited amounts of data storage (i.e. memory) otherwise the limit on available memory could further limit the range of computational tasks which the computer could solve. In order to achieve Turing universality a computer must have at least a minimum degree of complexity. A simple pocket calculator with a fixed set of functions, which cannot be programmed to alter its behaviour, is not a universal Turing computer. However, mobile phones, PCs, supercomputers and indeed most other computational devices in everyday use are Turing universal. Even the humble chip inside a chip-and-pin credit card is a universal computer, albeit a relatively slow one.

It has been proven formally that the information processing capabilities of the biomolecular machinery in living cells are Turing complete (Cardelli 2010). Indeed various sub-systems are themselves Turing complete computational mechanisms. Of course there are many differences between biological and silicon computation. Biological computers can store a gigabyte of information in less than a millionth of a cubic millimetre, they can perform huge numbers of computations in parallel at the same time, they are robust to failure, they are self-repairing, they are very efficient in their use of energy, and they can reproduce themselves. So our silicon computers have a long way to go. However, when it comes to the raw ability to do number crunching they are way ahead of biology, and your laptop is not going to be replaced by a blob of green slime any time soon.

The realization that molecular biology is performing universal computation highlights some deep connections between biology and computer science, and is reflected in the new cross-disciplinary field of research: synthetic biology (Royal Academy of Engineering 2009). The goal is to use the insights and techniques of engineering and computer science to understand the extraordinary complexity of living systems, and thereby allow living cells to be reprogrammed in order to modify their properties.

Synthetic biology is not just the preserve of professional scientists with multi-million-dollar research laboratories. There is an increasing movement of amateur groups, and even high-school students, experimenting with synthetic biology. Short sequences of DNA can be ordered over the web, simply by typing in the base-pair sequences into a web browser (along with credit-card payment) and samples of the corresponding DNA arrive by post a few days later. Currently the cost is around 30p per base pair, and falling with time. Note that such sites only deliver to bona fide users, and sequences are checked first for known pathogens to prevent nefarious activities. What if you don't know which sequences to use? Well, in 2003 Tom Knight at MIT introduced Biobricks, a standard registry of biological components (each consisting of a DNA sequence). The registry currently lists around 5000 items. There is an annual competition called iGEM (international Genetically Engineered Machine) for undergraduate teams around the world. An example iGEM entry is called 'E-chromi', and is a variant of the E-coli bacterium which has been modified to produce one of five different colours in the presence of an input signal such as an environmental toxin.

On a larger scale, Jay Keasling from the University of California, Berkeley, with support from the Gates Foundation, has created a yeast strain, using ten genes from three organisms, that is able to produce the anti-malarial drug artemisinin. This is an effective anti-malarial treatment that is traditionally extracted from plants and is very expensive. The new yeast could reduce the production cost by a factor of ten. Keasling also co-founded Amyris, which uses a similar approach to engineer yeast for making diesel fuel and high-grade chemicals.

Much of the work to date has focused on the use of existing genes having known functionality. To realize the full potential of synthetic biology, it will be essential to work in the reverse direction: for a given desired property of an organism, what is the corresponding DNA sequence that will give rise to that property? It is here that computer science has much to offer, and as an example we return to the topic that we started with: Turing patterns. A collaboration between Microsoft Research and the Department of Plant Sciences at Cambridge University aims to produce a variant of the bacterium E-coli which will form colonies that exhibit Turing patterns. Figure 4.11 shows a simple biomolecular circuit that should be capable of achieving this (Service 2011).

The cell can emit and also sense two chemical messengers, one acting as a short-range activator and the other as a long-range inhibitor, to implement the mechanism discussed earlier. The resulting balance of activation and inhibition determines the level of production of a protein

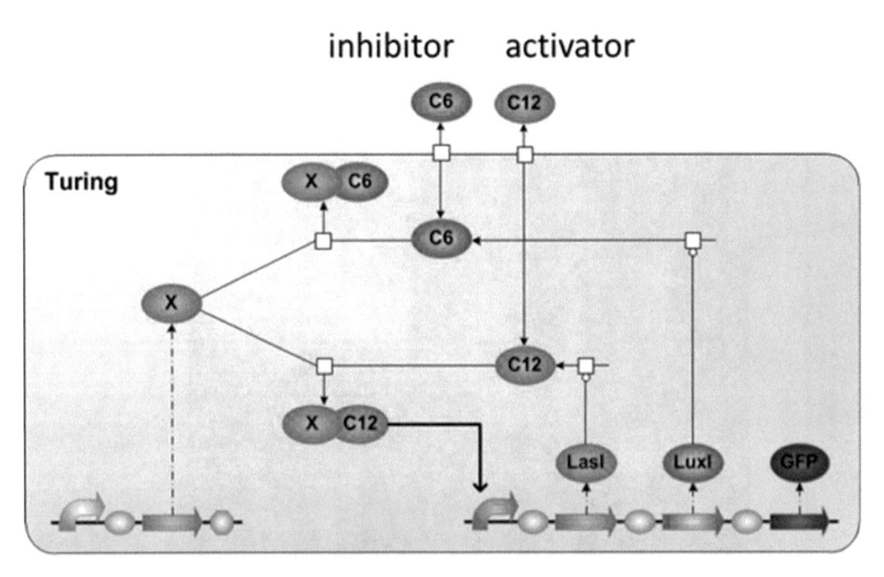

FIGURE 4.11 Biomolecular circuit for the production of Turing patterns in E-coli bacteria.

called GFP (green fluorescent protein) which can be detected by shining ultraviolet light on the cell. A description of the desired mechanism is then translated into computer code and fed into a piece of software called GEC ('genetic engineering of cells') (Pedersen and Phillips 2009) which designs a biomolecular system having the desired behaviour, drawing on a database of DNA components. A computer simulation of the behaviour of a colony of bacteria equipped with the corresponding DNA is shown in Figure 4.12.

Although still in its infancy, the field of synthetic biology has tremendous potential. Today we live in an unsustainable world in which the technologies we use to feed and clothe ourselves, and to provide materials for modern life, involve the one-way consumption of finite resources. We urgently need to move to a sustainable world based on renewable resources, and for the most part that will mean biologically based resources. Our ability to reprogramme biology, with the help of tools and insights from computer science and engineering, will be hugely valuable in that endeavour.

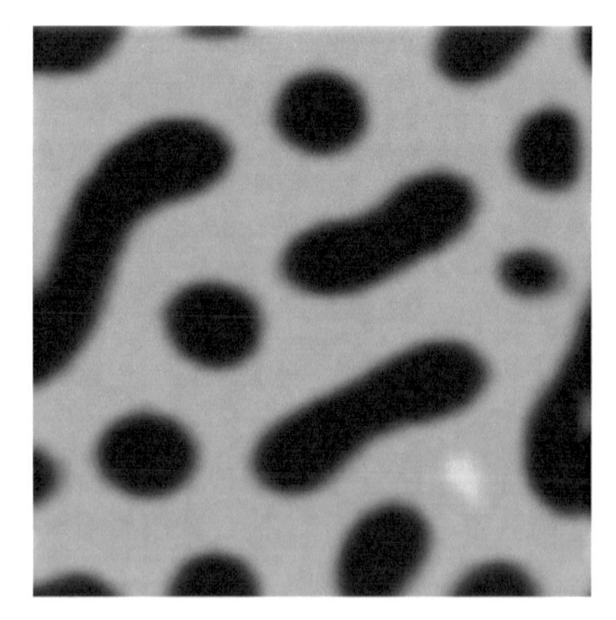

FIGURE 4.12 Simulation of cell development showing the emergence of a Turing pattern.

References

Bishop, C. M. 2006. *Pattern Recognition and Machine Learning*. New York: Springer.

Cardelli, L., and G. Zavattaro. 2010. 'Turing Universality of the Biochemical Ground Form', *Mathematical Structures in Computer Science* 20 (1): 45–73.

Hodges, A. 1992. *Alan Turing: The Enigma*. London: Vintage.

Kondo, S., and T. Miura. 2010. 'Reaction-Diffusion Model as a Framework for Understanding Biological Pattern Formation', *Science* 329: 1616–20.

Pedersen, M., and A. Phillips. 2009. 'Towards Programming Languages for Genetic Engineering of Living Cells', *J R Soc Interface* 6 Suppl. 4: 437–50.

Royal Academy of Engineering. 2009. 'Synthetic Biology: Scope, Applications, and Implications'. www.raeng.org.uk/synbio, accessed 3 March 2014.

Service, R. F. 2011. 'Coming Soon to a Lab Near You: Drag-and-Drop Virtual Worlds', *Science* 331: 669–71.

Shotton, J., A. W. Fitzgibbon, M. Cook, T. Sharp, M. Finocchio, R. Moore, A. Kipman and A. Blake. 2011. 'Efficient Human Pose Estimation from Single Depth Images', in *IEEE Conference on Computer Vision and Pattern Recognition.* Trier: Computer Science Bibliography. 1297–304.

Turing, A. M. 1950. 'Computing Machinery and Intelligence', *Mind* 59 (236): 433–60.

1952. 'The Chemical Basis of Morphogenesis', *Philosophical Transactions of the Royal Society of London. Series B, Biological Sciences* 237 (641): 37–72.

5 Life in conflict: soldier, surgeon, photographer, fly*

MARK DE ROND

> I drink not from mere joy in wine nor to scoff at faith – no, only to forget myself for a moment, that only do I want of intoxication, that alone.
>
> Omar Khayyam

Rarely have conflicts created such rich pickings for the media as those in Iraq and Afghanistan. A combination of smartphones, tablets, social networking and YouTube has generated a magnificent, even if occasionally grisly, assortment of snapshots and factual documentaries, blogs and autobiographies. Some have been controversial – think of Abu Ghraib – but mostly they have been flattering in espousing the heroism of ISAF forces. This is particularly true of images of human anguish beamed to our television screens by embedded photojournalists, a handful of whom, like Tim Hetherington and Rémi Ochlik, paid the ultimate price.

Despite unprecedented access, the ensuing reportage often overlooks the surreality of war by not being explicit about the conflicting experiences it generates for those involved first-hand. War is surreal for the paradoxes it mobilizes. Prominent among these (and quite aside from the surreality of the physical setting) are the want of community and camaraderie and yet the experience of competition and rivalry; the conflicting emotions of pleasure and guilt; sharp contrasts between a sense of meaning and futility. While petty by comparison, similar paradoxes may be found in the organizations that dominate much of our working lives. It may just be the case that they are easier to identify in contexts that are exceptionally austere, and where getting the wrong end of the stick kills.

* This chapter is a slightly extended version of an essay published under the title: 'Soldier, Surgeon, Photographer, Fly: Fieldwork beyond the Comfort Zone', published in *Strategic Organization* 13 (3): 256–62.

These conflicting experiences cannot easily be reconciled. Rather, those affected often have little choice but to reconcile themselves to them as best they can.

In this short contribution we move from the grotesque to the trivial by examining the experience of war from four vantage points: that of the soldier, the combat surgeon, the photojournalist and the ethnographer. I was encouraged to write from experience. When the opportunity arose to embed with a team of combat surgeons in mid-2011 in Afghanistan, I chucked in my lot lock, stock and barrel. I was given permission to take photographs – a privilege not even extended to those in charge – and returned with 1500 useable images. Preparations for my stint began in mid-2009 and involved eighteen months of interaction and negotiation followed by six weeks of pre-deployment training and six weeks in Afghanistan. (A tour of duty for UK surgeons is six weeks.)

The ensuing rumination is raw. I have not been back long enough to provide the depth of reflection that makes for good theorizing, and there will be plenty of exceptions to my inferences. So be it.

Soldier

Even if the surreality of war is never far from the surface in written accounts, it takes a bit of digging to find true gems. Here is one: Karl Marlantes, a former US marine and Rhodes Scholar, described his conflicting experiences as follows:

> I've jumped out of airplanes, climbed up cliff sides, raced cars, done drugs. I've never found anything comparable. Combat is the crack cocaine of all excitement highs – with crack cocaine costs . . . When I returned from the war I would wake up at night trying to understand how I, this person who did want to be a good and decent person and who really tried could at the same time love an activity that hurt people so much.
>
> (2012: 160, 68)

Marlantes' experience is not atypical. Jack Thompson, a veteran of several wars, refers to such experiences as 'combat highs': 'like getting an injection of morphine – you float around, laughing, joking, having a great time, totally oblivious to the dangers around you . . . Problems arise when you want another fix of combat, and another, and another and, before you know it, you're hooked.' Field Marshal Slim likewise described one of his personal kills as brutal but as giving him 'a feeling of the most intense satisfaction' (Grossman 2009: 236–7).

A US Army soldier, until recently deployed in the violent eastern Afghanistan's Korengal Valley, wrote this:

> [The] Scouts shot a man, and we did a gun run and somehow his leg came off. He was crawling around without a leg, and we were watching it on the ELRAS, the heat seeker. So you can see that his leg is gone, and the dude slowly crawled around, and stopped crawling, and we all fucking laughed . . . You think about it now and it's like, 'That was a fucking human being, you son of a bitch. You fucking crazy bastard, that was a human being you fucking killed. How can you fucking live with that? How can you fucking think that that was okay?' . . . That's the terrible thing of war. You do terrible things, and then you have to live with them afterwards. But you'd do them the same way if you had to go back. It's like an evil thing inside your body.
>
> (Hetherington 2010: 189–90)

Yet others describe their experience of killing as one of disgust. Says one: 'I dropped my weapon and cried . . . There was so much blood . . . I vomited . . . I felt remorse and shame. I can remember whispering foolishly "I'm sorry"' (Grossman 2009: 238).

Occasionally it is cowardice that produces guilt. As Anthony Loyd recalls in *My War Gone By I Miss It So*:

> As we walked to the edge of the hotel, our final cover before Snipers' Alley, [my Bosnian friend] told me that he would not run across to the other side, explaining that he 'never ran for those people' ['those people' meaning the Serbs] . . . I was in a dilemma. To run across alone would be unforgivably rude and craven . . . Yet to walk the stretch and risk being shot on the first day for the sake of politeness seemed equally stupid. The compromise was uncomfortable and a little ridiculous: as he strolled slowly across, face raised to [where the snipers were], I kept abreast of him walking sideways like a crab, first one way and then the other, hoping that anyone who wanted to shoot us would take out the easy target first.
>
> (Loyd 2000: 17)

War's surreality is occasioned by the contradictions of life on the front-line: the co-existence of meaningfulness and yet futility, protocol and compassion, trust and vigilance, adrenaline-fuelled pleasure and guilt; the appreciation of human significance but also its absurdity and mortality; an understanding of the value of life as enshrined in law and yet observing daily the carelessness and ease with which it is dispatched; the adrenaline rush of being shot at and killing in turn; the fascination with, yet horror of, mutilation, death and decay.

Surgeon

In Camp Bastion's hospital, to feel guilty typically means to feel callous for not caring more in the face of misery. As one of the surgeons told me at the time, the recent death of his Staffordshire terrier had affected him far more than anything he had seen here. It can also highlight inconsistencies in people's reactions to different types of casualties. As another put it to me:

> I remember . . . where someone died right in front of me, and I suddenly realized I didn't care. It didn't do anything to me. But then I saw a US marine come in who had been shot in the head and was now paralysed. That really did affect me. To see a young man, younger than me, barely old enough to drink and now being unable to move.

Surgeons and nurses used to talk of 'compassion fatigue' and the shame if not surprise that humanity can do such terrible things.

To treat major trauma effectively requires surgeons and anaesthetists to align their efforts in a context where the margin for error is small and the stakes literally a matter of life and death. Even if the social configuration around the operating table will vary with injury type and availability, all involved will have had plenty of practice working alongside different personalities and specialities. Thus, the anaesthetist will attempt to stabilize the patient's physiology while the general surgeon takes responsibility for the chest and belly. The limbs are left to orthopods and the face to a plastic surgeon. With a blast injury, casualties may require a double amputation, which means that the general surgeon may need to move into the belly first to clamp off the major vessels to the legs. But he can only do this after the anaesthetist is confident that the casualty is stable

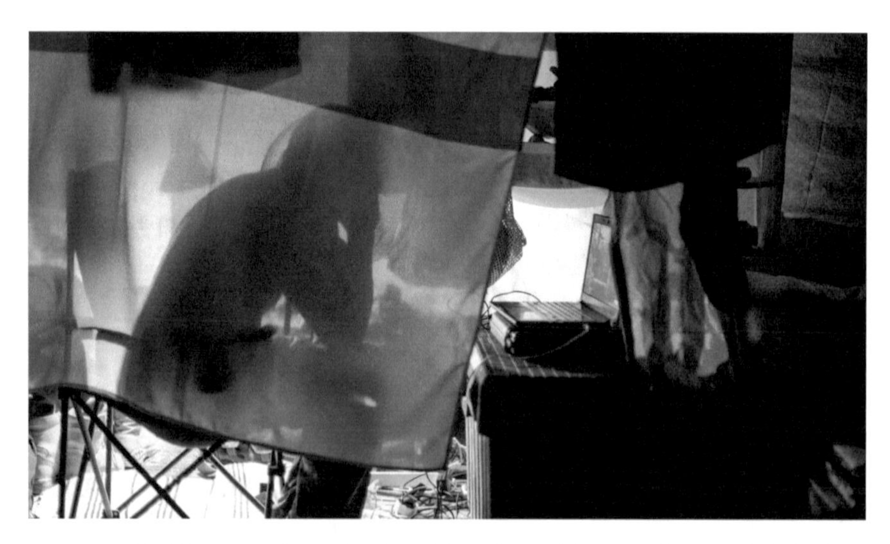

FIGURE 5.1 Camp Bastion, Afghanistan. July 2011. To try and find solace and privacy, one of the doctors draped flags and linens over clothes lines in a tent shared with seven other medical personnel.

physiologically. The scrub nurses will need to have the right equipment available at the right time; the orthopaedic and plastic surgeons will take their cue from the general surgeon as he does from them when it comes to unclamping the vessels. In other words, successful and safe treatment hinges on careful co-ordination.

Yet even such close co-operation does not rule out rivalry. For leave these surgeons with little or nothing to do work-wise and they may turn on each other instead. Unable to sit still, some begin to interfere in the affairs of others or to compete for work. As one of the surgeons admitted: 'He is fighting for work. I am fighting for work, each of us hoping the other will be late.' Sebastian Junger described the troops he embedded himself with as so bored on occasion that 'they prayed for contact [with the enemy] as farmers pray for rain' (Hetherington 2010: 15). Even when work is plentiful, surgeons may compete for the most interesting jobs.

As in Junger's Korengal Valley, in Camp Bastion's hospital periods of great intensity follow periods of boredom in which it is however impossible to relax or to put oneself to productive use (Figure 5.1); surgeons and warriors alike intentionally objectify casualties yet feel callous for not caring more than they do. It is here that the extremes of busyness and boredom, significance and futility can change rapidly and unpredictably, and shift the balance between altruism and selfishness, pleasure and guilt, the thrill of warfare and cowardice. 'In this kind of war', wrote McCullin, 'you are on a schizophrenic trip. You cannot equate what is going on with anything else in life . . . None of the real world judgments seem to apply. What's peace, what's war, what's dead, what's living, what's right, what's wrong? You don't know the answers' (2002: 100–1).

Photographer

The conflicting experiences of war also affect those who are first-hand witness to it through the viewfinder. Think of such stalwarts as Robert Capa and his grainy Omaha Beach images, of Eddie Adams' snap of the summary execution of a Vietcong by Saigon's chief of police, or of Don McCullin's picture of the mandolin player strumming happily over the dead body of a girl whose half-burnt house he had just ransacked – these images, alongside their private journals, leave very little to the imagination. Or consider a more contemporary example: Kevin Carter and his Pulitzer-Prize-winning photograph of a young girl stalked by a vulture in civil-war-torn Sudan. The photograph caused a sensation upon publication. Everyone wanted to know what happened to the girl. How did Carter intervene? While the *New York Times* ran an editorial explaining that the girl had made it to a feeding station 100 m away, the fact of the matter is that Carter didn't help. He could have easily picked her up and carried her there, and indeed this is what many of those around him suggested he should have done. The fact is, he didn't. At first Carter told people that he had chased the vulture away and then sat under a tree and cried. As the barrage of questions kept coming, he elaborated the story by claiming that he had seen the child get up and walk to the clinic. But it wasn't until after the award ceremony at Columbia University that Carter came clean with his editor in explaining that he had worked the situation, walked all around the child, working the scene from different angles. What he had really wanted was for the bird to flap its wings, he said. It would have made the photograph so much more dramatic. It was the co-existence of pride and shame, co-operation and rivalry that marked the drug-fuelled, rollercoaster career of Carter, and one that ended prematurely with his suicide in 1994. Fellow photojournalist, Don McCullin, upon hearing he had won the World Press Photo Award, wrote of his uneasiness of receiving a prize for depicting the suffering of others. Greg Marinovich described his inner turmoil at watching Zulus beat a Xhosa to death:

> 'Mlungu shoota!' one of the dozen killers exclaimed as they finally took note that I was taking pictures. The men sprang away, but within seconds they would surely realize that I was a defenseless witness to the murder. Fear swept over me. I prepared myself to do anything to survive: I thought of kicking the dead man and calling him a Xhosa dog. I was even prepared to spit on the corpse. I knew I would be capable of desecrating the body to survive.
>
> (Marinovich and Silva 2000: 16)

Fly (on the wall)

The lived experiences of the ethnographer are trivial by comparison. Yet even in my limited experience of war, the guilty pleasures of voyeurism contrasted sharply with my shame at not feeling able to reciprocate emotionally when faced with horrifying injuries or death (Figure 5.2). Days of comparative calm led me to hope for stiffer stuff, even if the more interesting injuries are invariably the more serious, meaning my growing appetite could be satisfied only at the expense of someone else. Gravitating to atrocity, I felt tainted by the suffering of others. My anxieties about being liked competed with concerns for the welfare of casualties. My scholarly ambitions jarred badly with the life-and-death scenarios faced by trauma surgeons, and served as a reminder of the cowardice of much contemporary scholarship. I was keenly aware of being able to beat my retreat at any time to grab something to eat or simply to recharge, without reason or explanation; an extravagance not parcelled out to surgeons and soldiers. I genuinely tried to do important work yet remained acutely aware that, even at the best of times, my academic output compares

FIGURE 5.2 Helmand, Afghanistan. July 2011. Mark de Rond resurfaces after having fired his first sniper rifle at a shooting range in Camp Bastion.

poorly to the most mundane of medical interventions. The headwork was exhausting. I returned home intolerant to bullshit.

> What are we to make of a creation in which the routine activity is for organisms to be tearing others apart with teeth of all types – biting, grinding flesh, plant stalks, bones between molars, pushing the pulp greedily down the gullet with delight, incorporating its essence into one's own organization, and then excreting with foul stench and gasses the residue. Everyone reaching out to incorporate others who are edible to him. The mosquitoes bloating themselves on blood, the maggots, the killer-bees attacking with fury and a demonism, sharks continuing to tear and swallow while their own innards are being torn out . . . The soberest conclusion that we could make about what has actually been

> taking place on the planet for about three billion years is that it is being turned into a vast pit of fertilizer.
>
> (Becker 2011: 282–3)

Bastion could be surreal in other ways too. One afternoon in early July, for example, a delivery arrived at the hospital: a carton box with a pair of mangled legs inside it. These had been sent to the hospital by well-meaning soldiers thinking that the limbs might arrive in time to be re-attached to the torso of their comrade, blown to bits by an improvised explosive device hours earlier. A little later on two badly burnt Afghans would arrive. One would be dead within the hour. The other would follow suit yet insisted that the hospital call a taxi to take him and his dead friend back to Korengal Valley. In the bed next to him was a fourteen-year-old boy. Classified as an insurgent he was denied access to books and toys, and watched over by a marine twice his size. A few beds over, a wild-eyed Afghan was blissfully unaware that he had been without a two-square-inch bit of skull for the past twenty-four hours. In the rush of things his skull-piece had been left behind in Kandahar. It arrived by plane in the early hours of the morning and was stuffed in our fridge, next to our chocolate snacks and zero-alcohol beer, awaiting re-attachment. In the background meanwhile there was the sound of a melody of a young boy victim singing.

Some reflections

While far less dramatic than the experience of surgeons or soldiers, the fieldwork affected me in ways I am still trying to understand. I remember trying to shed the Bastion experience by leaving behind almost all my clothes and shoes for fear of transporting the experience to the comforts back home. Paradoxically, I often felt ashamed for not feeling more emotional and needed reassuring that it was okay not to feel more deeply affected. Yet I also reminded myself with every new casualty that this wasn't my war and that, as surgeons would tell themselves too, we were not responsible for their suffering. I consciously forced myself into first-hand experience of blood and human flesh by mopping up gummy puddles after surgery. I could pray no longer. I had stopped dreaming.

None of it lasted. What has remained are a lingering impatience with all things trivial and a confusion about what matters and why. At no prior moment have I felt so disillusioned with what to me often feels like a publishing game of our own devising. If we took stock of the problems we have solved to date, and their consequences for humanity, would we have reason to be proud? Are there diminishing returns to methodological sophistication? Where are the *real* problems that should guide our research? We know the world is complex and that our knowledge of it is imprecise and incomplete, and so where is that point beyond which we profess to 'know' and, based on this, to act? When did we ever stop human suffering on such scales as witnessed in Iraq and Afghanistan – or on any scale for that matter? What did we ever do to stop the wars?

As for slipping into combat surgery, I hoped that an experience of the extraordinary might help unlock the ordinary. For the ordinary, or so it seems to me anyway, can be more difficult to decipher than its opposite; after all, it is so commonplace, so void of passion and sparkly bits. These suspicions are not far removed from Jean Cocteau's, who described his job as poet as consisting of 'placing those objects of the world which have become invisible due to the glue of habit in an unusual position which strikes the soul and gives them a tragic force'. In smoking out the contradictory experiences of war as a medic, I became increasingly sensitive as to parallel paradoxes in other forms of social organization – professional service firms, for example – even if the latter are likely to be subtler (and more mundane). While ethnography is a poor candidate for generalization of any sort, it seems right intuitively that forcing smart people to cohabit in an environment where there is a premium on getting it right will give rise to a similar set of conflicting experiences. For example, senior associates in a tax advisory firm co-operate on client projects yet simultaneously compete for partnership opportunities; partners in a law firm understand the importance of cross-selling yet compete for reputational effects; academics may appreciate the need for institution building yet are keenly aware too that their only real currency is their publication record. To do 'social' science is to live with the realization that much of what we do is impractical and easily dismissed as futile, and yet many of us plough on in the conviction that we can in fact better the world.

Some of the contradictions that characterize lived experience at the frontlines and inside the field hospital (fieldwork) mirror those of the organizational ethnographer (headwork). These contradictions cannot easily be reconciled. Instead, we must reconcile ourselves to them – in much the same way surgeons, photographers and soldiers do – by being explicit about their reality and legitimacy. This might help us understand why some break down in the face of suffering while others are much less affected, and why people can express their dissonance or grief in such very different ways. It helps one appreciate how surgeons can tire of both compassion and work, and can crack jokes at the expense of the sick yet be uncompromising in their care for them. Black humour alleviates where empathy fails. It is here that friend and foe, youthful and nostalgic, sceptic and compliant, profane and pious can co-exist, sharing a common priority never more sharply in focus as in a field hospital: a human being who, for his injuries, can no longer help himself but relies wholly and utterly on the generosity and capability of his peers. To that extent, Camp Bastion's surgeons struggle as we do too with what Jim March and Thierry Weil call the great conflicts of life: a predilection towards equality and modesty but also an urge to power and self-assertion; a commitment to rationality and the pursuit of self-interest but also a conception of duties and the pursuit of justice; a fascination with and yet aversion to killing, mutilation and death; a claim of human significance but so too an awareness of human absurdity and mortality (2005: 10). As Erich Fromm surmised, the essence of man is his paradoxical nature, the fact that he is half animal and half symbolic (Becker 2011: 27–8).

One of the luxuries afforded by a personal reflection is speculation, so here's my two cents: I would not be at all surprised if the management of these paradoxes is found to have consequences for psychological safety. For example, organizational teams where internal rivalry is actively discouraged, where people are forced to 'work as a team', and where it is all about the collective and not the individual, may also inhibit people speaking freely for fear of being castigated or considered 'not a team player'. This, in turn, may drive rivalry underground such that people begin to compete by belittling the work of others in private. Organizations that pride themselves on trust and camaraderie may discourage people from speaking out for fear of being thought unsupportive or of vigilant of the intentions of others. It is not hard to see how the way in which these paradoxes are managed may be related to perceived levels of psychological safety and ultimately, in the case of a medical unit, patient safety.

As I close, keenly aware of the sedate but comfortable foil that is Cambridge, I miss the adrenaline-fuelled world of combat surgery and the trough of sorrow and indignity into which we dip and by which we are stained. 'The human heart dares not stay away too long from that which hurt it most. There is a return journey to anguish that few of us are released from making', Lillian Smith observed. The ordinary no longer quenches one's thirst, and I am pained by the realization that the fly's feat is perhaps no different from that of the soldier, surgeon and photographer in that our pleasure may well be proportionate to another's pain.

References

Becker, E. 2011. *The Denial of Death.* London: Souvenir.

Grossman, D. 2009. *On Killing: The Psychological Cost of Learning to Kill in War And Society.* New York: Back Bay.

Hetherington, T. 2010. *Infidel.* London: Chris Boot.

Loyd, A. 2000. *My War Gone By I Miss It So.* London: Anchor.

March, J., and Thierry Weil. 2005. *On Leadership: A Short Course.* Oxford: Blackwell.

Marinovich, G., and Silva, J. 2000. *The Bang Bang Club: Snapshots from a Hidden War.* London: Basic.

Marlantes, K. 2012. *What It Is Like to Go to War.* London: Atlantic.

McCullin, D. 2002. *Unreasonable Behaviour.* London: Vintage.

Smith, L. 1949. *Killers of the Dream.* New York: Norton.

6 Life in the ancient world*

MICHAEL SCOTT

> Only the future is uncertain, the past is always changing.
>
> Anonymous

Not everything is quite as it seems

The year 2012 is – as it is becoming increasingly difficult to forget – an Olympic year. The Olympics have a long history, as we are often told, which stretches back to the world of ancient Greece. This link between ancient and modern was trumpeted again during 2012, not least at an exhibition entitled 'The Olympic Journey' at the Royal Albert Hall, which told the story of the Olympics from ancient Greece to the present day.[1]

And yet, just what picture of the ancient games, and thus the links with our modern Olympics, do we have in our heads? A quick quiz highlights the issues. Which of the following *did* form part of the ancient Olympics?

(a) the Olympic torch relay
(b) the Marathon race
(c) male athletes tying up their penises with string

Only the last is true of the ancient games. The torch relay was introduced by Hitler at the 1936 Berlin Games, and the Marathon race first became

* My sincerest thanks go to the Master and Vice Master of Darwin College for inviting me to speak and write for the Darwin College Lecture Series and to Janet Gibson for her unfailing efficiency and kindness in its organization. Thanks also go to Ioannis Galanakis for his thoughts on modern Greek approaches to life in ancient Greece, Clare Foster for her insights on 'swords and sandals' Hollywood films and to Paul Cartledge for his thoughts on the text as a whole. All remaining mistakes are, of course, my own.

[1] See www.olympic.org/culture-and-olympic-education-commission. Accessed 8 March 2014.

FIGURE 6.1 Recreation of the Temple of Zeus and Sanctuary at Olympia. © DAI – F. Adler, R. Borrmann, W. Dörpfeld, F. Graeber and P. Graef, *Die Baudenkmäler von Olympia*, Olympia II (Berlin 1892), Table 132.

part of the Olympics in 1896, at the inaugural modern games. Athletes tying up their penises with string, on the other hand, known as 'infibulation', 'ligaturing' or by its ancient name *kynodesme* ('dog-tying'), was a well-known feature of ancient athletics.[2] Its purpose is, however, unclear, with some scholars arguing that it was meant to help avoid unwanted erections, others simply to keep the penis out of the way when running, others that it was an issue of sexual attraction and others still an issue of modesty.

Nor are the differences between ancient and modern games just in the details. In fact I would argue that the entire context and tenor of the games has shifted substantially. The ancient Olympics were held in the middle of a religious sanctuary: they were fundamentally linked with worshipping the gods. Only Greeks could compete, indeed only men could compete. Married women were not even allowed to watch, in part because the athletes all competed naked.

Nor were the ancient Olympics a very pleasant experience (Figure 6.1). Imagine it – 40,000 or so Greeks in the middle of hot Greek summer,

[2] The original thesis on the subject is Dingwall 1925.

camped in tents around the sanctuary and stadium at Olympia, along with all the animals and other foodstuffs necessary to feed them. Imagine the animals being sacrificed, their throats slit, on some of the seventy-plus altars around the religious sanctuary at Olympia, the blood congealing in the heat as parts of their bodies were burnt to the gods and parts roasted for consumption, sometimes as many as 100 oxen at one go. Imagine the human and animal waste building up around the site from such a large crowd in a place with no drainage or sanitation. It is not for nothing that the Eleans, who ran the sanctuary, sacrificed to Zeus Apomuios – Zeus the 'Averter of Flies'. Athletes sweating, bleeding, sometimes dying; crowds pushing, orators, philosophers, historians, nutcases mouthing off at every opportunity, stallholders selling, politicians bickering. No wonder that the ancient sources talk of the difficulties of getting accommodation (even Plato had to bunk up in a stranger's tent), of getting transport away from the games afterwards, and there is even one famous ancient piece of advice that if you wanted to punish a slave, you sent him to the Olympics.[3]

Now, of course, there are similarities between the ancient and modern games, and that's the key point.[4] In making links between the ancient and modern Olympics, we have in fact actively shaped an image of the kind of ancient Olympics we want to be linked to, and ignored the rest. We have fashioned a picture of the ancient world in our minds, which corresponds to what we like to emphasize in our world today. This is by no means a new game; in fact, it is one we have never stopped playing since the time of the ancient world itself. In this chapter, I examine *how* and *why*, over the centuries between them and us, we have gone about formulating a picture of what life was like in the ancient Greek world, analyse the kinds of answers we have come up with, and ask how our picture might continue to change in the future.

[3] Tents: Aelian 4.9; Sweaty noisy conditions: Epictetus 1.6.26; Impossibility of finding transportation: Lucian *Peregrinus* 35; Punishment for slave: Aelian 14.18; Zeus Apomuios: Pausanias 5.14.1. For the nature of sport in the ancient world: Crowther 2004; Spivey 2004; Crowther 2007. For discussions of the site of ancient Olympia see: Finley and Pleket 1976; Raschke 1988; Scott 2010; Stuttard 2012. For a forthcoming discussion of the nature of sporting venues in ancient Greece: Scott forthcoming.

[4] As former UK Culture Minister Tessa Jowell recently put it: 'while the games themselves may have altered dramatically, many of the values and ideals that underpin the Olympic movement have remained unchanged'. In Goff and Simpson 2011: 204.

Taking an interest

I pick up the story at the moment in which Europe, long fascinated by the surviving literature of the ancient world, began to become interested in what was left of the physical remains of ancient Greece. It might well come as a surprise that we haven't always been interested in ancient ruins. Yet if you think that Greece, after the break-up of the Roman Empire, was subsumed as an unimportant and difficult-to-get-to backwater in the Byzantine Empire, and given that it became part of the Eastern Orthodox Church after the Schism of 1054 and thus unappealing to the Catholics of Western Europe, followed by its absorption into the Ottoman Empire after 1453; it is easy to see why few in Western Europe cared about visiting and about what 'what was there' could tell them about life in ancient Greece.

And yet, by the fifteenth century, things were also beginning to change. The Humanist movement in Italy fostered an interest once again not only in the surviving texts of ancient Greece, but also its coins and inscriptions, and gradually in the remains of the place itself. The famous Italian merchant Cyriac of Ancona travelled all over the Mediterranean in the second half of the fifteenth century, trying to map literature on to geography.

What did these early explorers seek and what picture did they come back with? Cyriac's vision of Greece, informed by ancient literature, was as a place full of heroes, legends, powerful gods, civic duty and philosophy, something expressed well also in Raphael's slightly later 'School of Athens' painting (1511) in the Apostolic Palace of the Vatican, depicting Greek philosophers at conversation. But crucially, what Cyriac and his contemporaries thought could be gained from this physical investigation was the evidence for a picture of an ancient world which the modern would do well to imitate. The image of life in ancient Greece as a goal to strive for had begun to gather momentum.

It is not a long step from imagining life in ancient Greece as a model for the present to wanting to associate yourself with its surviving remnants, and in turn to owning them. Through the fifteenth to seventeenth centuries, artists directly imported ancient figures into their paintings and kings and the aristocracy of European nations, particularly England and

France, started to acquire ancient Greek art. The ruins of ancient Greece morphed from object from which lessons could be learnt, to symbol of a gentleman's 'connoisseurship', to pawn in international power-politics among European royalty. Not in all cases with the resulting respect for the objects you might have thought. One antique column drum, owned by a certain Mr James Theobald during this period, was apparently used as a roller for his bowling green at his home in Berkshire.[5]

As a result of this development in ancient art connoisseurship, by the late seventeenth century, the focus of interest in this idealized conception of ancient Greece was also changing. Up to this point, the 'model' from ancient Greece had often been that of Sparta, the city of men with über-six-packs, long hair and a 'spartan' approach to life and art.[6] But as the arts – their understanding and collection – came to be a more and more honourable pursuit in the sixteenth and seventeenth centuries, the focus for a model in ancient Greece began to shift from 'spartan Sparta' to 'artistic Athens'.[7] That is to say, as the modern world's interests changed (in part thanks to the re-appearance of ancient Greek ruins), so too did the modern world's focus on what was best in ancient Greece, and, as a result, the emphasis of the portrait painted of life in the ancient Greek world. The ancient Greeks were no longer only sharpening their swords to kill, but also sharpening their tools to carve.

Yet, while the notion of Greece – and particularly Athens – as an ideal was developing alongside the acquisition of bits of its ruins, just what picture of the landscape, architecture and 'shape' of ancient Greece was being formulated? The irony is that, apart from depictions in the surviving literary sources, no one had a clue as to what ancient Greece had really looked like. Despite pilfering many of its surviving pieces of art, there was still no sustained interest in investigating, recording or understanding what was left of its landscape. In 1554, a century after

[5] For example, the collection by the Earl of Arundel (1586–1646), known as the Arundel Marbles, contained the first Greek artefacts to reach England. At his death, the collection contained thirty-seven statues, 128 busts and 250 inscriptions, most of which eventually passed to the Ashmolean Museum. Cf. Stonemann 2010: 48–52. Theobald: Constantine 2011: 10.

[6] James Harrington in *The Commonwealth of Oceana* (1656) called on Oliver Cromwell to be a new Lycurgus, the mythical law-giver of Sparta in the seventh century BC.

[7] Cf. the famous comparison of 'spartan' Sparta and 'artistic' Athens in Thucydides' *History of the Peloponnesian War* 1.10.

FIGURE 6.2 Nicolas Poussin, 'Body of Phocion Being Carried out of Athens', 1648. © National Museum of Cardiff.

Cyriac, Martin Kraus, a Greek scholar at Tübingen in Germany wrote to Theodoros Zigomalas, a Greek in Constantinople, to enquire whether Athens still existed or had been replaced by a fishing village.[8] Nicolas Poussin's 1648 painting of the 'Body of Phocion Being Carried out of Athens' records a key historical moment in Athens' history as presented in the literary sources (Figure 6.2). But his vision of ancient Athens is almost entirely made up from his imagination based on his interpretation of the ancient literature. Indeed so flexible was the public's impression of ancient Athens that Guillet de Saint-George, in the second half of the seventeenth century, was able to publish a highly successful book about Athens based on a fictional visit to the city by his fictional brother![9]

Greece, it seems, was an ideal because, to a large extent, it was a blank canvas, onto which people could impose their own ideas picked up through their own interpretations of the literary sources. Moreover, what

[8] Etienne and Etienne 1992. 33.
[9] Guillet de Saint-George, *Athènes ancienne et nouvelle* (1675; published under the name of his brother La Guilletière).

FIGURE 6.3 Anon. [J. Carrey?], sketch drawing of part of the West Pediment of Parthenon, 1674, from H. Omont, *Athènes au XVIIe siècle*. Paris 1898.

is fascinating is that for many, this notion of ancient Greece as an ideal demanded the corresponding belief that there was nothing left of it to find even if they went looking (you could not have the 'ideal', it seems, without the nostalgia for its current 'loss'), a thought which extended to assuming the ignorance of the modern Greeks about their glorious ancestry. Guillet de Saint-George's book also poked fun at the way in which the modern Greeks actually played up to European expectations by feigning particular ignorance about ancient culture in their presence.

In the last decades of the seventeenth century, however, this interest in art began to encourage people systematically to fill in the blank canvas and record what was left of ancient Greece. The Capuchin monks, originally sent in by the French to convert the Orthodox Greeks, started making the maps of ancient Athens. Painters attached to the French Ambassador to the Ottomans, the Marquis de Nointel, made the first sketches of the Parthenon and its sculpture (Figure 6.3).[10] Jacob Spon and George Wheler, undertook the first detailed investigations and drawings of surviving monuments around Greece, in the process having to convince the

[10] Drawings attributed to Jacques Carrey: Constantine 2011: 12.

Turkish soldiers that they were not spies (measuring old bits of stone the Turks found very suspicious) and more often than not bribe them.[11]

By the end of the seventeenth century, the picture of life in ancient Greece had been established as an ideal, its ruins as valuable commodities, but the nature of that picture was still largely a creation of European imaginations and ancient literature, with increasing pockets of more detailed knowledge of surviving ruins, particularly in the now fashionable-to-admire Athens. But even those surviving pieces were themselves – sometimes brutally – reshaped to fit with modern European morality. The Duc de Mazarin, taking over Richelieu's collection in the Palais Royal in 1670, went around with a hammer knocking off the genitals of ancient statues so affronted was he by their nudity. This sense of effrontery at, and attempt to disguise, the seeming sexual liberation of the ancient world would continue for a long time. Visit any collection of ancient Greek art today, and you will see any number of ancient statues that have had a fig leaf added to their nether regions, not irregularly on the direct orders of the Pope himself.[12]

Nostalgia, admiration and reclamation

By the beginning of the eighteenth century, the ability to fictionalize completely a tour of ancient and modern Greece was no longer viable: Guillet de Saint-George's fiction was exposed to great outrage in London by Spon and Wheler. In its place, the new craze was Homer, and particularly the ability to identify the locations in Homer's text in the Greek landscape, and in so doing, to link life in ancient and modern Greece back together because, it was felt, Homer's genius had originally emerged from a contemplation of that same landscape.[13] The pleasure of reading Homer

[11] During his visit, 12.5 kg of coffee secured Spon access to the Acropolis. The published account and drawings are in J. Spon, *Voyage d'Italie, de Dalmatie, de Grèce et du Levant fait aux années 1675 et 1676 par Jacob Spon et George Wheler* I–II. Amsterdam, 1679.

[12] Constantine 2011: 6–12.

[13] Cf. the publication, first in 1769, then properly in 1771 and again in 1775, of Robert Wood's *Essay on the Original Genius and Writings of Homer with a Comparative View of the Ancient and Present State of the Troade*. Wood had written the text much earlier in his life, but did not come to publish it until later due to his intervening career as a politician.

was immeasurably improved by reading it in the locations he was talking about. But at the same time, that process affirmed a change in the perception of the realities of ancient Greece. As Goethe later said 'we stopped seeing in Homer's poems a strained and inflated world of fabulous heroes and saw instead the true reflection of a primitive reality'.[14]

In part linked to this new-found desire to read Homer in situ, European knowledge of ancient Greek art and architecture was increasing in leaps and bounds. But it was also due to recent greater freedom to travel in this region, and, indeed, a social emphasis on its desirability. The Grand Tour was now a fixed part of any gentleman's upbringing and the Society of the Dilettanti was formed in 1734 in England, for aristocrats who had visited Italy. The Society went on to fund numerous voyages to ancient lands and their resulting publications, like those of Richard Chandler (1764–1766), and, critically, those of James 'Athenian' Stuart and Nicholas Revett, who published detailed plans of many ancient Greek temples and buildings in several volumes after 1762 (Figure 6.4). Copying these structures became *de rigueur*: the monument of Lysikrates, which was to be found in the kitchen-garden of the Capuchin monks who had drawn the first map of ancient Athens, was copied as far away as Philadelphia in the United States, where it crowned the Merchant's Exchange building, completed in the early nineteenth century.[15]

Yet Greek architecture was not the only focus of interest, particularly for German scholars including Johann Winckelmann. The Germans had only recently become interested in ancient Greece (as a result of the new craze for Homer and the Greek landscape), and Winckelmann now led the way with his construction of the first history of ancient Greek art. Indeed so keen on Greece was Winckelmann that he famously claimed in a letter to his friend 'I have never wanted anything so passionately as this [the chance to visit Greece]. I wouldn't mind losing a finger, in fact I wouldn't mind losing my balls for such a chance of getting to see those countries.'[16] Despite numerous opportunities, however, he never went.

[14] Johann Wolfgang von Goethe, *Aus meinem Leben: Dichtung und Wahrheit* (1811–33) IX.537–8.

[15] Expeditions of Richard Chandler were published in 1775 as *Travels in Asia Minor* and, in 1776, *Travels in Greece*. James Stuart and Nicholas Revett, *Antiquities of Athens*, 1762–1816. Cf. Stonemann 2010: 127.

[16] *Briefe an Bianconi* (Letters to Bianconi), II.69.

FIGURE 6.4 James Stuart and Nicholas Revett, *Antiquities of Athens*, vol. 1. Chapter 4. Plate 3: Lysikrates monument.

Whatever his protestations, the first scholar to deliver a history of Greek art did so from Rome.

Despite the distance, Winckelmann's influence on the nature of Greek art, and thus the nature of the ancient Greek world in which it had been created, was fundamental. Using statues like the Apollo Belvedere (Figure 6.5), the same statue to which Winckelmann was forced to witness the Pope ordering the attachment of a fig leaf in 1759, Winckelmann argued that the production of great art was tied to political and social freedom, but he also trumpeted the 'noble simplicity and calm grandeur' of Greek sculpture.[17] That highlighting of simplicity and grandeur (made possible by liberty) chimed with the goal of the modern European Enlightenment: the promotion of nature and reason. As a result, this new vision of life in ancient Greece was able to continue smoothly as a 'new' ideal for a new modern society.

Did this continued – indeed increased – idealization of life in ancient Greece still demand a counterpart expectation of ignorance among the modern Greeks as it had in the seventeenth century? The issue was hotly debated. Edward Gibbon continued the more traditional line that the modern Greeks were pale imitations of their ancestors: 'it would not be easy in the country of Plato and Demosthenes to find a reader, or a copy, of their works. The Athenians walk with supine indifference among the glorious ruins of antiquity.'[18] Yet, some felt the exact opposite. Pierre Augustin Guys, in 1771, published his account of a comparison between ancient and modern Greek values, arguing that the modern Greeks had much of the ancient spirit. That meant both a natural simplicity and occasionally a tendency to be 'artful, vain and not very scrupulous observers of their oaths'.[19] Guys's work was political in two ways. First, it threw stones at the overly fussy and degenerate culture of aristocratic Western

[17] Cf. *Briefe an Bianconi* III.62–3. Winckelmann's major works: *Gedanken über die Nachahmung der griechischen Werke in Malerei und Bildhauerkunst* (1755); *Geschichte der Kunst des Alterthums* (1764); *Monumenti antichi inediti* (1767–8). The key passages are collected in Irwin 1972. Cf. Etienne and Etienne 1992: 45–6; Whitley 2001: 22; Constantine 2011: 93–142.

[18] Cf. Constantine 2011: 142.

[19] P. A. Guys, *Voyage litteraire de la Grèce ou Lettres sur les Grecs anciens et moderns, avec un parallele de leurs moeurs* (1771; English edition) I.25. For a similar argument of connection between the ancient and modern Greeks, see Johann Hermann von Riedesel, *Remarques d'un voyageur moderne au Levant* (1773).

FIGURE 6.5 The 'Apollo Belvedere' statue.

Europe: the sentence 'such were the simplicity and good sense of ancient manners, we are far removed from them today' (I.54) was removed in the English translation of his work. But, his work also intimated that the modern Greeks, currently under Turkish occupation, were merely dormant and had it in them to return to the glories of their ancient ancestors. Guys boasted Catherine the Great of Russia as one of his avid readers, and it is no surprise that she, in 1770, had a hand in inciting the Greeks to rebel against the Turks. But the dismal failure of that rebellion had the unfortunate result of encouraging many to side with Gibbon that the modern Greeks simply weren't up to their ancestors' standard.

By the end of the eighteenth century, the answer to the question what was life like in ancient Greece was perhaps even more complex than it had been a century earlier. On the one hand, it was all about noble simplicity, calm grandeur, naturalism and reason, which could be understood for some only from within the landscape of Greece with Homer in hand, and, for others, from afar through its art. That noble simplicity, made possible by ancient liberty, in turn, was portrayed either as a goal for the modern Europe Enlightenment to head towards, or as a stick with which to punish those countries which were, now, thought too 'over-sophisticated', as well as cover to hint for greater political and social freedom in European society. Yet, at the same time, depending on how you read the character of the modern Greeks, ancient Greece was also either totally lost or, in fact, balanced on the edge of a comeback.

Collecting, independence and archaeology

The nineteenth century was to prove no less complex in its attitudes towards ancient Greece and its image of ancient Greek life. By the beginning of the nineteenth century, in part because more and more people were travelling to Greece, seeing its current state of desolation, hearing about the lacklustre performance of the modern Greeks compared to the ancient Greek ideal, the overall impression of life in ancient Greece began to take on a foreign, distant note. As Leslie Hartley later put it in his novel *The Go-Between*: 'the past is a foreign country – they do things differently there' (Hartley 1953: 1). Art at the time seemed to capture this distant, foreign, otherness of the ancient world. Paintings in the early nineteenth century of Greek landscapes are famously bathed in a haze of golden light, which is the antithesis of the often extraordinarily clear outlines and quality of light natural to Greece. Even the artistic vision of ancient Greece was separated from the viewer by mist.[20]

The irony is that, just as, for some, life in ancient Greece seemed more distant, for others, knowledge and attachment to the Greek landscape was also becoming stronger than ever. Edward Clarke, travelling in Greece in 1800–1, claimed that 'Epidauria is a region as easily to be visited as Derbyshire'. Indeed, it wasn't just that visiting Greece was easier than it had been, it was that every part of the country mirrored that of England: Thessaly was the Yorkshire Dales, Ligourio was Cheltenham.[21] At the same time, the first expeditions intent not on drawing sculpture or architecture, but on mapping the Greek landscape, set out. British Army Officer William Martin Leake, in his travels in Greece 1805–7 and 1809–10, established the location of many of the famed, yet so far lost, sites of ancient mainland Greece.[22]

And whether ancient Greece was like a foreign country or whether its modern landscape was like Cheltenham, Europe still wanted to own what remained. In 1801–3, Lord Elgin packed 100 cases of Parthenon sculpture to ship to London. In 1812, the sculptures of the Temple of Aphaia on Aegina were sold to King Ludwig of Bavaria. Nor was this only happening in Greece. The French at this time carried off the Luxor obelisk to stand in the Place de la Concorde in Paris. However we view these actions today, at the time, the main players were adamant of their moral right to do this. 'Antiquity', said Capitaine de Verninac Saint-Maur, responsible for taking the Egyptian obelisk, 'is a garden that belongs by natural right to those who cultivate its fruits'.[23] Yet in bringing these sculptures to Europe, another debate was ignited about the nature of Greek art, and ancient Greece itself. Seeing the Parthenon sculptures up

[20] E.g. Charles Lock Eastlake, 'Byron's Dream', 1829, and Hugh William Williams, 'The Temples of Jupiter Panhellenius (Aphaea) in Aegina', 1820.

[21] Edward Daniell Clarke, *Travels*, I–VII (4th edn 1818); cf. also William Otter, *Life and Remains of Edward Daniell Clarke* (1825).

[22] William Martin Leake, *The Topography of Athens* (1821); *Travels in the Morea*, I–III (1830); *Travels in Northern Greece* (1835); *Peloponnesiaca* (1846).

[23] Quoted in Stonemann 2010: 165. For discussion on the Parthenon marbles and other pieces taken out of Greece at this time: St Claire 1984.

close, artists and scholars, expecting the 'noble simplicity' of Winckelmann, were astounded by the detailed rendition of human reality. Greek art was no longer about a mathematical canon, but about real human beings. So fundamental was this change that many, at first, felt driven to deny the Parthenon sculptures entirely as a work of Greek art. Richard Knight, on behalf of the Society of the Dilettanti, argued that Elgin had misidentified as Greek what were in fact Roman sculptures from the time of Hadrian. It was not until 1816 that Elgin was able to convince the British Museum of their veracity and worth.

The Greek War of Independence (1821–9) called Western Europe's bluff on their claim to idealize ancient Greece. Some of Europe put their lives where their admiration led them. The Germans sent 300 fighters to help the Greeks, more than any other nation.[24] Lord Byron, who not only wrote many poems extolling the virtues of ancient Greece and vociferously attacked Lord Elgin for his taking of the Parthenon sculpture (cf. *Childe Harold*, 1812), but also recreated famous ancient acts like Leander's nightly swim across the Bosphorus from Sestos to Abydos to reach his beloved, died at Missolonghi while acting as a Greek freedom fighter.

Without doubt the war also helped close the gap once again between a 'distant' ancient Greece and the modern present. For the modern Greeks fighting in the war, it was time to lay claim to their ancestry. As Alexandros Ypsilantis, one of the early Greek leaders in the war, exclaimed in 1821: 'Brave and valiant Greeks, let us remember the ancient freedom of Greece, the battles of Marathon and Thermopylae, let us fight on the tombs of our ancestors who fell for the sake of *our* freedom.'[25] It is interesting to note that too, in the decades after the war, ancient Greek names became increasingly popular for modern Greeks, as they sought to strengthen their connection with their past.[26] But for Europe, the War of Independence also ignited an ever-stronger bond between modern Europe and ancient Greece as expressed through the recognition of the

[24] Morris 1994: 25.

[25] Quoted in Etienne and Etienne 1992: 85. Cf. Dakin 1972.

[26] Greek names: I. Galanakis 2012 (personal communication). See the influential work by Konstantinos Paparrigopoulos, *History of the Greek Nation* (1860–77), which outlined the connection between ancient and nineteenth-century Greece.

Greek impact on the European present. John Stuart Mill, in 1846, is famed as saying that:

> the true ancestors of the European nations are not those from whose blood they are sprung, but those from whom they derive the richest portion of their inheritance. The battle of Marathon, even as an event in British history, is more important than the battle of Hastings. If the issue of that day had been different, the Britons and the Saxons might still have been wandering in the woods.[27]

The rest of the nineteenth century bore the fruit of this belief in, and determination to foster European closeness to ancient Greece. As Greece itself was tied closer into Europe with the establishment of Prince Otto of Bavaria as King of the Hellenes in 1833, so too progressed the development of the scientific study of Greece's ancient remains, along with the need to protect them, conducted by both Greeks and other European nations. The Acropolis was declared the first archaeological site in 1834. In 1836, the first archaeological law was passed in Greece regulating what could and could not be sold, supplemented by a second in 1899. In 1837, the Greek archaeological society was formed. In 1846, the French School in Athens was set up. In 1874 Heinrich Schliemann published, or rather adorned his wife with, the results of his excavation at Troy and in 1876 began to dig at Mycenae. In 1874, too, the German Archaeological Institute was founded, and in 1875, the first of the big digs commenced at Olympia, which was entirely buried underneath metres of alluvial sediment, followed by that of Delphi in 1892, which was also buried deep under a modern settlement. Both of these excavations, along with many others, represented enormous financial investments for the excavating nation, testament to their deep connection to ancient Greece.[28]

What picture of life in ancient Greece did these advances – the birth of archaeology itself as an academic discipline – have to offer? On the one hand, of course, apart from simply exposing key sites, it brought huge advances in understanding their nature. Careful examination of temples, for example, during the second half of the nineteenth century proved that

[27] Published in Mill's 1846 review of George Grotte's *History of Greece* (1846–56). Cf. Whitley 2001: 30.

[28] In 1886 the British School at Athens was founded and many major academic journals of archaeology began to circulate during that decade. Dyson 2006: 159.

FIGURE 6.6 The – almost entirely – fifth-century BC Acropolis as it is seen today. © Michael Scott.

they had been painted – a fact that astonished a world so used to 'marble' classical ruins, and who had now to confront an ancient world looking like one of the most kitsch places on Earth.[29] Yet, with the exception of Schliemann's famous discoveries at Troy and Mycenae, as well as those of the British archaeologist Arthur Evans at Knossos on Crete, the focus was also most often on creating a picture of the ancient world in what was considered its heyday: the fifth century BC, the acme of the classical period, the time of the Athenian democracy, empire and the Parthenon. The Acropolis, for example, in being turned into an archaeological site, was stripped of its complex architectural history back down to the fifth century BC (Figure 6.6), and excavations at Delphi showed little interest in anything before the Archaic Period levels in the first instance.[30]

The development of archaeology in Greece, and the resultant display of (a particular era) of 'real' ancient Greece, also did little to kill off the imagined ideal of ancient Greece. Indeed that ideal was busy morphing once again. In part due to the age of industrialization in the second half

[29] The first extensive discussion was J. J. Hittorf, *De l'Architecture polychrome chez les Grecs* (1830); cf. Stonemann 2010: 253. For colour in ancient sculpture see Brinkmann 2003; Panzanelli 2008.

[30] Radet 1992; Stonemann 2010: 242. For Schliemann, his critics and his impact: Etienne and Etienne 1992: 110; Stonemann 2010: 265–80; Constantine 2011: 200–5.

FIGURE 6.7 Lord F. Leighton, 'Greek Girls Playing at Ball', 1889. © Dick Institute, Kilmarnock.

of the nineteenth century, the concept of Greece as a timeless, rarified, intellectual world was set aside in European art, and it instead became a spectacular world of endless holiday, of which Lord Leighton's 1889 'Greek Girls Playing at Ball' is an excellent example (Figure 6.7). Ancient Greece was now a place of escape from the realities of the European world.[31] The very forces of idealization that had provoked an interest in the surviving remains of the ancient Greek world, and led ultimately to the birth of archaeology, now, not only continued unchecked by what those remains had to say, but offered, very much, an 'alternative' picture.[32]

In the same period, and in complete contrast to the artistic view of ancient Greece, the view of ancient Greek life through its surviving texts was also reframed in particular by the work of Friedrich Nietzsche, who published his work on Greek tragedy in the 1870s, offering a fundamental revision of ancient Greece as a place of both light and dark, of constructive

[31] Tsigakou 1981: 77.

[32] Morris argues that, from the 1870s, the study of the archaeology of Greece was systematically absorbed into Departments of Classics and cut off from the broader study of world archaeology, in order to protect the tradition of idealistic Hellenism, which it was beginning to threaten (a case of keep your friends close and your enemies closer!): Morris 1994: 11.

and destructive forces, of the power of the gods Apollo and Dionysos.[33] For Nietzsche, the idealization of Greece was a symptom of ignorance about the 'real' antiquity, a place that, if seen, would horrify the modern world.

Thus as the modern world emerged from the Victorian age into the twentieth century, life in ancient Greece was simultaneously a Butlins holiday camp, a world of light and tragic cruelty, an everlasting ideal, as well as a place whose ruins (most often fifth-century, and preferably democratic, ruins) were increasingly coming into focus and, as a result, a world that could be studied, catalogued and grasped.

Vikings, Polynesians, diversity and difference

Ancient Greece continued to play its role as inspirational ideal well into the twentieth century. Buses in London during the First World War carried excerpts from the speeches of Thucydides as inspirational adverts – this was only possible because ancient Greek literature was so embedded in the education system. In modern Greece too, poets like Cavafy and Seferis were taking inspiration from ancient Greece, and using ancient myths to describe current political and social conditions.[34] The Greeks were also increasingly combining their ancient heritage with the other crucial period of their past: Byzantine Greece. In fact, there is still a term in Greek: *hellenorthodoxos politismos* (Greek-Orthodox civilization), which evokes a blending of ancient Greece and Orthodox Christianity, and which some in Greece continue to call on today as an ideal.[35]

As a result of all this, it was felt, in the early twentieth century, that somehow the modern European world was now bound tighter than ever before to ancient Greece. In a Berlin bookshop not long ago I came across a 1921 volume edited by Richard Livingstone on *The Legacy of Greece.* The preface makes the European close relationship with ancient Greece clear:

> In spite of many differences, no age has had closer affinities with ancient Greece than our own. History does not repeat itself. Yet if the twentieth

[33] Shanks 1996: 173; Constantine 2011: 212.
[34] On Cavafy and Seferis see: Jusdanis 1987; Beaton 1990.
[35] I. Galanakis 2012 (personal communication). Cf. also Hamilakis 2007.

> century searched through its past for its nearest spiritual kin, it is in the fifth and following centuries BC that they would be found. Again and again as we study Greek thought, behind the veil woven by time and distance, the face that meets us is our own, younger, with fewer lines and wrinkles on its features and with more definite and deliberate purpose in its eyes. For these reasons, we are today in a position, as no other age has been, to understand Ancient Greece, to learn the lessons it teaches, and in studying the ideals and fortunes of men with whom we have so much in common, to gain a fuller power of understanding and estimating of our own.[36]

One of the ironies of this close, family, bond between early twentieth-century Europe (or its academics at least) and ancient Greece is that, at exactly the same time, the art world, which had for so long fanned the flames of idealistic interpretations of Greece, had, at the beginning of the twentieth century, began to dispense with it. Post-Impressionist artists such as Cézanne labelled its imitation pernicious, and the Dada movement, in the 1920s, suggesting 'letting Laocoön' – one of the most famous ancient Hellenistic sculptures known through Roman copy (Figure 6.8) – 'and his children rest after their 1000 year long struggle with that fine sausage of a serpent'.[37]

Yet it is also important to realize that this sense of affinity felt in intellectual circles between ancient and modern Greece did not stifle the doubts of Nietzsche and others surrounding the dual nature of the ancient Greek world: its light and dark, creativity and cruelty. In an article in Livingstone's volume entitled 'The Value of Greece to the Future of the World', Gilbert Murray tackled this Janus-like quality. On the one hand, he argued, the language of Greek poetry had an 'austere beauty' because the people were 'habitually toned to a higher level of intensity and nobility than ours'. On the other, he argues, the Greeks were separated by a thin and precarious interval from the savage. 'Scratch a Russian', Murray continued, 'and you find a wild Tartar. Scratch an ancient Greek, and you hit, no doubt, on a very primitive and formidable being, somewhere between a Viking and a Polynesian.'[38]

36 Livingstone 1921: Preface.

37 Etienne and Etienne 1992: 115. Cf. also the continuing disappointment felt when seeing Athens for the first time, in the 1920s: Byron 1926.

38 Murray 1921: 11, 15.

FIGURE 6.8 The Laocoön statue.

The dual savage and cultured nature of ancient Greece has not been, of course, the only fascination for academic scholarship during the twentieth and twenty-first centuries. It would be impossible here to list all the developments across the fields of classical study and how they have impacted upon our impression of life in ancient Greece.[39] But it is crucial to note that, while many of these developments and their resulting impressions have come about through new critical theory or new technology, just as many of them have been motivated principally by pressure,

[39] For a good review, see Snodgrass 1987; Whitley 2001: 35–57; Dyson 2006: 152–250.

as in previous centuries, from wider historical events as well as changing social and political attitudes.[40] Hitler's championing of the competitive elitist aspects of ancient Greece ensured that money was available for the Germans to dig out the stadium at ancient Olympia, which had hitherto been left unexcavated. In contrast, post-Second World War, with Europe exhausted, America came to the fore in archaeology in Europe, with the study of ancient Greece exploding with a particular emphasis on ancient Greece's credentials for freedom and democracy, with a resulting swing back from Olympia to Athens: in 1952–1956, the Americans paid to rebuild the Stoa of Attalus in the Athenian agora. For the modern Greeks, however, post-Second World War politics led to the expression, for example in Greek poetry, of the ancient Aegean landscape as no longer a place of blue skies and sunshine, but as a place of oppressive desolation.[41] In contrast again, Russian interest in ancient Greece, dating back to Catherine the Great, was preconditioned for much of the twentieth century by the politics of the Soviet Union, with its scholars choosing to emphasize an ancient Greece which was a pre-capitalist social system relying on slave modes of production.[42] The development of the European Union project after 1958 encouraged many to emphasize the panhellenic, 'united' aspects of ancient Greece, and particularly its sanctuaries such as Delphi, Olympia and Delos, as forerunners of the modern European movement. When UNESCO made Delphi a World Cultural Heritage site in 1987 it cited as an explicit reason 'Delphi's enduring ability to bring people together'.[43] And, for the modern Greeks, the particular problem of having to live in among the ruins, despite their pride and high passion for their ancient ancestry, has also occasionally pushed some to see it as a burden.[44] This came to the fore particularly during the heavy summer

[40] Other often-cited social and political changes: the emancipation of women leading not only to female scholars and archaeologists, but also to a fundamental rethink about the position of women in the ancient world and an active search for them in the historical record. Equally, the gradual acceptance of homosexuality has led to a more open discussion of the evidence for ancient sexuality, much of which had previously been hidden or ignored. Cf. Dover 1978; Dynes and Donaldson 1992.

[41] Kotsovilli 2009. [42] Cf. Dyson 2006: 214.

[43] See http://whc.unesco.org/en/list/393/documents. Accessed 8 March 2014. Ian Morris argues that the study of ancient Greece, over the last two centuries, has in reality been all about understanding our own Europeanness: Morris 1994: 8.

[44] Alexander the Great, for example, was voted the greatest Greek of all time on Greek television in 2009. But a German magazine was sued in the Greek courts for

fires of 2007, when decisions had to be made as to whether to protect ancient ruins rather than modern neighbourhoods.[45]

Conclusion

Our vision of life in ancient Greece has thus always been closely linked to the events, attitudes and needs of our own world. As a result, our impression of ancient Greek life, and the level of importance it has for modern society, has not just changed over time, but has also been dynamic and multiple at any one time. It has morphed, been contested and circled around time and again as a fictional ideal and primitive reality, distant foreign land and family member, place of noble simplicity and savage cruelty to name but a few. As the archaeologist Michael Shanks once put it: 'The Classical past does not reveal itself in its essential character, it has to be worked for. This leads us to the question: what sort of Classical past do we want?'[46] As a result of this constant tussle over the nature of the ancient Greek world, and its constant evocation in different guises as an ideal (and anti-ideal) in modern society, ancient Greece has a power, and punches way above its weight in comparison to other great periods of world history. The question becomes, will that continue in the future?

The argument has been made many times, even by Classicists themselves, that we cannot rely on ancient Greece to have the same kind of stake, if indeed any stake, in the twenty-first and coming centuries as it has done in the past,[47] and many readers will be aware of the continuing fight for the relevance of studying the ancient past in school curriculums as well as at university level. That war is certainly not over, but I do think battles are being won. For sure, putting excerpts of Thucydides on buses as inspirational messages would be somewhat odd now, and I doubt anyone would claim, as Livingstone did, that we are a mirror reflection of ancient Greece, or agree with John Stuart Mill that Marathon is more important than Hastings. But, on the other hand, the evidence is,

representing an ancient Greek goddess raising her finger to Europe: Associated Press, 29 November 2011.

[45] Cf. the discussion in Hamilakis 2007; Loukaki 2008.

[46] Shanks 1996: 118. Or as J. Porter puts it: 'Hellenism... is a relation between a particular past, itself differently imagined over time and therefore not very particular at all, and an ever-changing present.' Porter 2009: 8.

[47] Cf. Morris 1994: 44; Shanks 1996: 108.

I think, that the study of the ancient world continues to have relevance, and more importantly, to thrive and captivate. Numerous individuals, societies, associations and projects, for example, are working hard at grass-roots level helping with the often very popular re-introduction of ancient languages into schools across the country. Despite some universities narrowing their ancient world study options, others are actually increasing their provision, and numbers of students applying to read Classics-related subjects at university in the UK are healthy.

In wider society too, ancient Greek themes continue to gain prominence. Many have noted the apparent popularity of 'swords and sandals' epics in the cinemas in the last decades. This rise may indicate not so much a 're-awakening' of interest in ancient history, but rather the potential the ancient world has to provide an evocative and exciting 'blank canvas' for Hollywood script-writing and cross-genre fertilization (video-game potential, for example, is now equally if not more important than the film itself in the marketing plan for these projects).[48] This manipulation of the ancient Greek world in modern cinema represents, on the one hand, nothing new in the way the ancient world has been used (and abused) in the past. And on the other hand, it also has the useful effect of creating a knock-on effect of interest in the ancient world in other genres and audiences. It cannot be unconnected that there has been a rise in the number of television documentaries about the ancient world in recent years, or that, for example, Vera Wang's 'ancient Greece' wedding dress has apparently become one of the most popular in her collection, or that Karl Lagerfeld decided to dedicate the 2011 Pirelli calendar to Greek myth (modestly comparing himself to Homer when he commented that 'what Homer did with the pen, I did with the camera lens'). Such a wide distribution of films 'inspired' by ancient Greece has also meant that the culture of the ancient world has spread to new parts of the modern world as a useable comparative ideal. When I was teaching in Rio in 2010, the police there were being hailed in the mainstream press as Spartan heroes (and people knew what they were talking about because of the film *300*).[49] All

[48] Cf. Berger 2011: 12; Foster 2012. For discussion of ancient Rome in cinema: Wyke 1997.

[49] Cf. in America, where marines have often been compared to ancient Spartans: e.g. Warren 2005.

this, combined of course with the current world economic crisis and the Greek economy's position as a litmus test of the wider-world financial crisis, means that we find ourselves reading more and more articles comparing ancient ideals and modern realities.

The cycle of interest in, and debate over, life in ancient Greece continues. None of this should make us complacent. The ancient Greek world continuing as a useful springboard for inspiration should not be confused with a re-awakening of passion for history, or a continuing desire to better understand the past. But such use and abuse of the ancient world has simultaneously encouraged academics to engage both more critically in the subject of reception of the classical world, and thus to at least ensure our awareness of the processes of use and abuse, as well as to engage more actively in the public debate over the nature of the ancient world.[50] And that is surely key: if interest in life in ancient Greece is to continue in a useful and constructive form in the longer term, it will only be because the argument for its continuing intellectual, social, political and cultural relevance continues to be forcefully made and won both in academic, educational and most importantly, public arenas.

References

Beaton, R. 1990. *George Seferis.* Bristol: Bristol Classical Press.

Berger, R. 2011. 'The Art (and Arc) of Game Writing', *Written By, The Magazine of the Writers Guild of America* November/December: 10–12, 68–70.

Biddiss, M., and M. Wyke, eds. 1999. *The Uses and Abuses of Antiquity.* Berne: Peter Lang.

Brinkmann, V. 2003. *Die Polychromie der archaischen und frühklassischen Skulptur.* Munich: Biering and Brinkmann.

Byron, R. 1926. *Europe in the Looking-Glass: Reflections of a Motor Drive from Grimsby to Athens.* London: Routledge.

Constantine, D. 2011. *In the Footsteps of the Gods: Travellers to Greece and the Quest for the Hellenic Ideal.* London: Taurus.

Crowther, N. B. 2004. *Athletika: Studies on the Olympic Games and Greek Athletics.* Hildesheim: Weidmann.

[50] Classical reception has seen a huge growth in scholarship in the last two decades: e.g. Biddiss and Wyke 1999.

2007. *Sport in Ancient Times.* Westport, CT: Praeger.
Dakin, D. 1972. *The Unification of Greece, 1770–1923.* London: Benn.
Dingwall, E. J. 1925. *Male Infibulation.* London: Bale and Danielsson.
Dover, K. 1978. *Greek Homosexuality.* London: Duckworth.
Dynes, W. R., and S. Donaldson, eds. 1992. *Homosexuality in the Ancient World.* New York: Garland.
Dyson, S. 2006. *In Pursuit of Ancient Pasts: A History of Classical Archaeology in the 19th and 20th Centuries.* New Haven, CT: Yale University Press.
Etienne, R., and F. Etienne. 1992. *The Search for Ancient Greece.* London: Thames and Hudson.
Finley, M. I., and H. W. Pleket. 1976. *The Olympic Games: The First 1,000 Years.* London: Chatto and Windus.
Foster, C. 2012. 'Screenwriting from a Practical Perspective: Some Myths Debunked', paper delivered to Centre for Research in Arts, Social Sciences and Humanities (CRASSH), University of Cambridge, 23 January 2012.
Goff, B., and M. Simpson, eds. 2011. *Thinking the Olympics: The Classical Tradition and the Modern Games.* Bristol: Bristol Classical Press.
Hamilakis, Y. 2007. *The Nation and Its Ruins: Antiquity, Archaeology, and National Imagination in Greece.* Oxford: Oxford University Press.
Hartley, L. 1953. *The Go-Between.* London: Hamish Hamilton.
Irwin, D., ed. 1972. *Winckelmann: Writings on Art.* London: Phaidon.
Jusdanis, G. 1987. *The Poetics of Cavafy: Textuality, Eroticism, History.* Princeton, NJ: Princeton University Press.
Kotsovilli, E. 2009. 'The Dark Side of the Sun: Aegean Islands as Places of Exile, Desolation and Death in Post-World War II Greek History', in G. Deligiannakis and I. Galanakis, eds., *The Aegean and Its Cultures.* Oxford: British Archaeological Reports. 139–44.
Livingstone, R., ed. 1921. *The Legacy of Greece.* Oxford: Clarendon.
Loukaki, A. 2008. *Living Ruins, Value Conflicts.* Aldershot: Ashgate.
Morris, I. 1994. 'Archaeologies of Greece', in I. Morris, ed., *Classical Greece: Ancient Histories and Modern Archaeologies.* Cambridge: Cambridge University Press. 8–47.
Murray, G. 1921. 'The Value of Greece to the Future of the World', in R. Livingstone, ed., *The Legacy of Greece.* Oxford: Clarendon. 1–24.
Panzanelli, R., ed. 2008. *The Color of Life: Polychromy in Sculpture; From Antiquity to the Present.* Los Angeles: John Paul Getty Museum and the Getty Research Institute.
Porter, J. 2009. 'Hellenism and Modernity', in G. Boys-Stone, B. Graziosi and P. Vasunia, eds., *The Oxford Handbook of Hellenic Studies.* Oxford: Oxford University Press. 7–18.

Radet, G. 1992. 'La Grande Fouille vue par un contemporain', in O. Picard, ed., *La Redécouverte de Delphes*. Paris: Boccard. 144–8.
Raschke, W. J. 1988. *The Archaeology of the Olympics*. Madison: University of Wisconsin Press.
Scott, M. C. 2010. *Delphi and Olympia: The Spatial Politics of Panhellenism in the Archaic and Classical Periods*. Cambridge: Cambridge University Press.
forthcoming. 'The Spatial Indeterminacy and Social Life of Athletic Facilities', in P. Christesen and D. Kyle, eds., *A Companion to Sport and Spectacle in the Ancient World*. Oxford: Wiley-Blackwell.
Shanks, M. 1996. *Classical Archaeology of Greece: Experiences of the Discipline*. London: Routledge.
Snodgrass, A. M. 1987. *An Archaeology of Greece: The Present State and Future Scope of a Discipline*. Berkeley: University of California Press.
Spivey, N. 2004. *The Ancient Olympics*. Oxford: Oxford University Press.
St Claire, W. 1984. *Lord Elgin and the Marbles*. 2nd edn. Oxford: Oxford University Press.
Stonemann, R. 2010. *Land of Lost Gods: The Search for Classical Greece*. London: Tauris Parke.
Stuttard, D. 2012. *Power Games: Ritual and Rivalry at the Ancient Greek Olympics*. London: British Museum.
Tsigakou, F.-M. 1981. *The Rediscovery of Greece: Travellers and Painters of the Romantic Era*. New York: Thames and Hudson.
Warren, J. 2005. *American Spartans: The US Marines, a Combat History from Iwo Jima to Iraq*. New York: Free Press.
Whitley, J. 2001. *The Archaeology of Ancient Greece*. Cambridge: Cambridge University Press.
Wyke, M. 1997. *Projecting the Past: Ancient Rome, Cinema, and History*. New York: Routledge.

Further reading

Angelomatis-Tsougarakis, H. 1990. *The Eve of the Greek Revival: British Travellers' Perceptions of Early Nineteenth-Century Greece*. London: Routledge.
Beaton, R., and D. Ricks, eds. 2009. *The Making of Modern Greece: Nationalism, Romanticism and the Uses of the Past (1797–1896)*. Farnham: Ashgate.
Borst, W. 1948. *Lord Byron's First Pilgrimage*. New Haven, CT: Yale University Press.

Colin, J. 1981. *Cyriaque d'Ancône: humaniste, grand voyageur, et fondateur de la science archéologique.* Paris: Maloine.

Damaskos, D., and D. Plantzos, eds. 2008. *A Singular Antiquity: Archaeology and Hellenic Identity in Twentieth-Century Greece.* Athens: Benaki Museum.

Detienne, M. 2007. *The Greeks and Us: A Comparative Anthropology of Ancient Greece.* Cambridge: Polity.

Hardwick, L., and C. Stray, eds. 2008. *A Companion to Classical Receptions.* Oxford: Wiley-Blackwell.

Goldhill, S. 2011. *Victorian Culture and Classical Antiquity: Art, Opera, Fiction, and the Proclamation of Modernity.* Princeton, NJ: Princeton University Press.

Miller, H., ed. 1972. *Greece through the Ages.* London: Funk and Wagnalls.

Mount, F. 2010. *Full Circle: How the Classical World Came Back to Us.* London: Simon and Schuster.

Soros, S., ed. 2006. *James 'Athenian' Stuart: The Rediscovery of Antiquity.* New Haven, CT: Yale University Press.

Taplin, O. 1989. *Greek Fire.* London: Jonathan Cape.

Valavanis, P., ed. 2007. *Great Moments in Greek Archaeology.* Athens: Kapon.

7 Life in ruins[*]

ROBERT MACFARLANE

I begin with a thought-experiment. Imagine, to start with, the disappearance of humans from the Earth. We are eliminated, in this scenario, not by means of a nuclear war or other pan-planetary catastrophe, but rather by a *Homo sapiens*-specific virus, which results in the swift deletion of our species but leaves our built environment intact, and the ecologies of which we are part undisturbed other than by our absence. Imagine, in fact, for the purposes of finessing the counter-factual, that the planet's last human perished at some point in the late winter of 2012. What then would happen to 'the world without us', in Alan Weisman's phrase (Weisman 2007: 5)? Or rather – for the more precise purposes of this thought-experiment – what would happen to the city of Cambridge without us? How would the city alter over the weeks, months, years, decades and centuries following its abandonment?

Allow me to hypothecate a set of futures for this post-human Cambridge. First and fastest come the hungry fungi, even at that cold time of year. In kitchens and dining rooms, moulds bloom on food left out on sideboards and worktops, spreading their mycelial nets of grey, yellow and green. With no one to chase it away, dust settles in the windless interiors of lounges and bedrooms. Decay has started, but so in its way has stasis. In March, a late spell of winter is cast. Water freezes in pipes, and then a thaw brings floods: ceilings crash down, walls fatten,

* Thanks are due for various reasons to Bill Adams, Guy Cuthbertson, Ruth Dewhirst, Brian Dillon, Richard Emeny, Abby Graham, Leo Mellor, Jules Pretty, Rosemary Vellender, Christopher Woodward and the Edward Thomas Estate. In the course of the essay I have drawn in particular on David Skilton's discussion of Macaulay's 'New Zealander' in Skilton 2004; on Will Viney's various fascinating discussions of lapsedness, ruin, waste and profit at http://narratingwaste.wordpress.com, accessed 9 March 2014; and on Dillon 2011.

soffit boards rupture. That summer, fire follows: lightning strikes and gas explosions, leaving cratered holes and husked houses. By August, rosebay willowherb – also known as bomb-weed, because it thrives on carbon-rich soil – flowers on these blackened sites like a pink floral fire.

Out on the streets in the first couple of years, the situation is volatile. Asphalt and paving have been cracked by freeze–thaw cycles; soil has blown into the fissures and plants have taken root. Sycamore seedlings prise up paving slabs and shift granite kerb edges, bindweed laces the spokes of rusting bicycles, jays bury acorns by the thousand in open ground. Aggressive species such as buddleia proliferate on facades and walls, their roots powerful enough to crack bricks when thirsting down for water. Buildings and thoroughfares are slowly torn apart by flora; weeds disassemble the city into tipsy chaos.

After five years, various guerrilla ecologies are well established. There are successful escapees from domestic gardens: snowberries, cotoneaster, conifers. Japanese knotweed and Japanese anemones run riot. The water in the city's outdoor swimming pools has thickened with leaf debris, which mulches darkly down; the first stage in the creation of tiny fens. Blackthorn marches from the hedges, suckering rapidly along, and elm works out in small thickets, keeping its head low (for Dutch Elm disease is still present). A soil cap deepens on the tarmac, low in nutrients and therefore highly bio-diverse, on which clover, grass, and wildflowers including orchids thrive.

Within five years, pairs of peregrines are breeding in the chalk pits of south Cambridge and on the towers of the university buildings, drawn by the increased prey and the absence of human interference. Owl populations boom, enjoying the availability of nest-sites in the ruins and the surge in mice numbers. Rabbit and fox populations fluctuate in complex relation to one another. Packs of wild dogs become the top mammal predators.

With the city's flood defences unmanaged, sluices and bridges block with debris, and the river begins to explore the possible extents of its domain. Coe Fen and Lammas Land flood: good news for little egrets and herons and for phragmites reeds. Chetti's warblers sing out of the vast stands of pink Himalayan balsam that bully the rest of the riverbank flora. Without agricultural nitrate run-off, the Cam becomes clear and oligotrophic, its pike more abundant and larger in size. The American crayfish completes its invasion of the waterways: mink and otter, no longer depleted by traps and cars, rise in number.

Over the course of a century, on the drier ground, grass gives way to thistle, which gives way to a scrub of hawthorn and elder, which gives way to a forest that grows extensively through the city, comprising sycamores, ash and oak, beeches, hazel and lime: a return of sorts to the pre-Atlantic period wildwood that once covered the south-east of England. On Parker's Piece, the open parkland in the city's centre, the grass first rises high and then falls back as silver birch and a heathland emerge.

Within 500 years nature's reclamation project is almost complete. Little that is recognizable of today's Cambridge is still visible. The city has been re-wilded. The famous buildings are mostly rubble, enjungled like Mayan ruins. The open spaces are forest or fen, browsed by large herbivores – Konick ponies and Highland cattle, escapees from local conservation initiatives – which keep glades open between the trees.

Of course, human residue remains, even that long after we have vanished. What will survive of us is love, wrote Philip Larkin, but actually what will survive of us is plastic. When archaeologists from other planets come to analyse the late-Anthropocene, our middens will be filled not with clam-shells and nut-husks, like those of our Mesolithic forebears, but with washing-up liquid bottles and ice-cream tubs. This is how my counter-factual ends: on a summer's day 1000 years hence, the warm wind trundles an empty plastic Coke bottle past what was once Great St Mary's Church, chasing it between trees and saplings, until at last the bottle gets ensnared in weeds, hard up against the pale and prostrate masonry of King's College Chapel.

Ruinism

Ecologically speaking, ruins offer niches for species: their combination of shelter and exposure, and their broken material textures, provide ideal footholds for weeds and wildflowers. Culturally speaking, ruins also offer niches for narrative: their disrupted structures, and their resonant allusions to collapsed pasts and dreamed-of futures, provide ideal footholds for writers and artists, who have for centuries now been drawn to ruins as sites peculiarly generative of story and of trope.

It is possible, indeed, to construct a long and near-continuous cultural history of what might be called 'ruinism' – by which I mean the art and literature of ruins – from classical literature through to the present day, with clear peaks of interest during the Renaissance, Romanticism, and then through the late nineteenth, twentieth and twenty-first centuries. Within that ruinist tradition, there has been an enduring preoccupation with nature's resurgence in a context of human wreckage – the idea of 'life in ruins', in both senses. This preoccupation has often taken the form of an apocalyptic pastoral in which – as in this essay's opening thought-experiment – ruins induce green-minded and at times troublingly misanthropic futurological fantasies about humanity's large-scale depletion and nature's large-scale return. The roots of this apocalyptic pastoral go back to the early seventh century BC, to the Book of Zephaniah, which relishingly warns that after the destruction of Nineveh 'the desert owl and the screech owl shall lodge on its capitals, the raven croak on its thresholds'. All such scenarios arise from the fact that, as Georg Simmel observed in his 1911 essay 'The Ruin', a ruin is an inorganic object sliding towards an organic state: these scenarios imaginatively follow that slide forwards to the point that artifice is all but abolished and organicism all but triumphant.

This essay is interested in the ways in which writers and artists since the late nineteenth century have imagined the future ruination of the human world, and how they have conjured the place of nature – life – within those ruins. It is consciously fragmentary in its engagements with numerous different works, rather than braced by a single arching thesis, but those fragments are also tendrilled through by certain continuous preoccupations. As well as exploring ideas of life in ruin, and the work of particular modern ruinist artists, works and texts, I want also to consider some of the conceptual paradoxes that aggregate around ruinist art, as well as the uses and capacities of future ruinism. Which is to say – to explore how speculative art about the ruins of the future might have been thought able to help us to visualize possible futures for ourselves, and therefore perhaps the better both to inhabit our own time, and to choose wisely between our various available prospects.

FIGURE 7.1 Gustave Doré, 'The New Zealander', from Blanchard Jerrold, *London: A Pilgrimage* (London: Grant and Co., 1872).

Among the best-known images of modern ruinism is Gustave Doré's 1872 engraving, entitled 'The New Zealander' (Figure 7.1). Thomas Babington Macaulay had first imagined the figure of the New Zealander in an 1840 review essay warning against Whig presumptions of the glorious continuity of English Protestantism. The Catholic Church, Macaulay concluded, 'may still exist' far in the future, when 'some traveller from New Zealand shall, in the midst of a vast solitude, take his stand on a broken arch of London Bridge to sketch the ruins of St. Paul's' (Macaulay 1840: 258). The rhetorical afterlives of Macaulay's New Zealander were prolific. By January 1865, the New Zealander had become so ubiquitous that *Punch* magazine issued a parodic proclamation banning its use. Henceforth, the editorial declared, 'it shall not be lawful for any journalist, essayist, magazine-writer, penny-a-liner, poetaster, criticaster, public speaker, lecturer, Lord Rector, Member of Parliament, novelist, or dramatist' to make use of this 'threadbare' and 'hackneyed' figure (Anon. 1865: 9). Nevertheless, seven years after *Punch*'s proscription, Doré included 'The New Zealander' in the set of 180 engravings he published to illustrate Blanchard Jerrold's *London: A Pilgrimage*, and Doré's image has since itself become iconic to the point of cliché. There sits the New Zealander with his sketchbook, looking across to Commercial Wharf: symbol of the city and its former life-bloods of finance and trade, now redundant on the burst banks of the Thames. Above the wharf rise ruins – and up to its capitols surges life in the form of the trees.

It is worth noting the voyeuristic aspect to Doré's image, for it shares this aspect with much of what might tendentiously be called 'romantic ruinism': there sits the witness, safe on his side of the river. He is no calamity-racked refugee: well clothed and well equipped, the ruins prompt him to cogitation, rather than challenging him to survival. And we, of course, are second-order voyeurs, watching the watcher – and ourselves able also to wallow safely in the spectacular pathos of the scene. Voyeurism is one of the two most frequent criticisms levelled at ruinist art and literature; the other is that of nostalgia or conservatism. According to this criticism, actual ruins draw the viewer's imagination always backwards in time, and the future ruin is often made emblematic – as here in Doré – of state institutions which, by being mourned in advance, are implicitly affirmed in the present. Certainly, ruins – either real or forecast – all allude to a function that is no longer fulfilled. The ruined cathedral has ceased to offer a space for worship, the ruined wharf has ceased to facilitate nation-building trade. The material incompleteness of a ruin can provoke a supplementary impulse in the viewer. We 'supply the missing pieces from [our] own imagination', as Christopher Woodward has put it (Woodward 2001: 15). It is for these reasons among others that the nostalgic-conservative criticism is made – the suggestion that the ruinist imagination of both artist and audience is inevitably passive-regressive.

There are limits to this criticism, however, and over the course of the nineteenth and twentieth centuries, ruins have come increasingly to be figured culturally not as nostalgic sites, but rather as complicatedly and diagnostically forward-looking. In fiction, sculpture, painting and film, the anticipation of a ruinous end has frequently served as a narrative means by which, or more precisely from which, we might return and make better sense of our freshly estranged present. Because future ruins allude to an end but are not quite it – matter and evidence have survived, traces are readable – they possess a peculiar power in terms of relaying the present. Again and again, imagined ruins have been pressed into cultural service, used both to diagnose and to warn.

Enjunglement and crystallization

Over the later nineteenth and twentieth centuries, ruinism tightened as a cultural fascination; in the late twentieth and early twenty-first centuries, it has approached an obsession and has therefore also often condensed as kitsch. We live now in an era of widespread ruination, and so we live also an era of widespread rumination on ruination. As Brian Dillon notes a fine short essay on the recent cultural history of decay, late modernity has experienced 'what appears to be a distinct flourishing – in the realms of global events, popular culture and the work of visual artists – of images of catastrophe and decay' (Dillon 2011: 10). One such late modern flourishing might be traced from its origin in Max Ernst's vast painting 'Europe after the Rain', thought to have been completed in 1942. The rain of Ernst's title is of course the rain of bombs delivered by Allied and Axis planes upon the cities of Europe, and the ruin of his canvas is part-masonry, part-ossuary: the humanoid figures, erotically tendrilled with lianas, seem to have been smothered rather than salved by the return of vegetation. The natural life that has returned to this debris is neither hopeful nor benevolent; it has collaborated in the extinction of the human presence, or at least its mutation into the hybrid birdmen and tree-people that can be seen emerging here and there in the painting (such portmanteau post-humans recur in the ruinist fantasies of Japanese anime directors, most notably Hayao Miyazaki and his followers, including Kei-ichi Sugiyama, whose *Origin* (2007) is set in a reforested future Earth, and features Ernst-inspired dryads and aged tree-men).

Ernst's art turns up as a detail in J. G. Ballard's 1962 novel *The Drowned World*, which is set in a future in which climate change has resulted in massive sea-level rise and the enjunglement of great swathes of the globe. On the wall of a ruined apartment in an overgrown city, writes Ballard, the canvas of 'one of Max Ernst's self-devouring phantasmagoric jungles screamed silently to itself'; one futurological work of ruinism is seen to have predicted the scenario contained within another (Ballard 2008: 29). In 1966 Ballard published the partner-piece novel to *The Drowned World*, called *The Crystal World*, the mass-market paperback edition of which wore another of Ernst's jungular paintings as its cover. The book's conceit is audacious: deep in the jungles of Central Africa, the world begins to crystallize. From its African epicentre, the crystallization moves outwards, converting the jungle and its inhabitants into a bejewelled dream-forest, in which crocodiles encased in glittering second skins lurch down the river, and pythons with huge gemstone eyes rear and strike lapidary poses.

Ballard's crystallizations themselves recur in a recent example of ruinist art, Roger Hiorns' 2008 sculpture/installation-piece, 'Seizure'. Hiorns arranged for 75,000 l of copper sulphate solution to be pumped into an abandoned and disintegrating council flat in a low-rise late-modernist housing development in Elephant and Castle, London, where it crystallized upon every available surface (Figure 7.2). It is a troubling, even menacing precipitation, a spiky modernist super-mould that suggests a decay first accelerated then halted: the witchily beautiful bedizening of a confined living area in a troubled part of a city. It is a work that allows no easy reading or soft voyeurism: it is too ocularly and conceptually prickly for that. The 'Seizure' of its title invokes at once a physical spasm, the chemical action of precipitation, and seizure in the sense of a compulsory

FIGURE 7.2 Detail from Roger Hiorns' 'Seizure' (2008): view of the crystallized interior of the Harper Road flat. Photograph copyright Nick Cobbing. 'Seizure' was commissioned by Artangel and the Jerwood Charitable Foundation, and supported by the National Lottery through Arts Council England, in association with Channel 4. Image reproduced by permission of Artangel.

purchase order or bailiff's visit to evict a household unable to pay its bills: the work was contemporary in its engagement with the consequences of financial crisis.

Late-modern ruinism has proliferated, of course, because late-modern ruins have proliferated. As Ernst suggests – and as Leo Mellor has documented in a fine recent book on the subject, *Reading the Ruins* – the Second World War made an actual ruinscape of much of Europe. During the Cold War, nuclear conflict proposed to the imagination an apocalypse that was both sudden and plausible: a day's laying waste of a planet that had taken 3 billion years of life to develop. As the Cold War threat receded, its brutalist infrastructures became redundant, and fell first to abandonment and then to ruin. And although the nuclear threat has itself receded, climate change now nourishes 'further ruinous fantasies and apprehensions' (Dillon 2011: 10). In the realms of architecture and urban planning, infrastructure obsolescence is now more rapid: as Woodward observes, 'destruction' is now 'of a different speed and scale... we construct more, and bigger, buildings than ever before – and abandon them more quickly than ever before' (Woodward 2011: 18). Financial crises leave uncompleted building projects; drastic urbanization continues worldwide – for many reasons, our present is increasingly littered with debris of the futures that our pasts have envisaged.

The *locus classicus* of our contemporary fascination with life in ruins is Detroit, the Motor City, which, after enjoying early twentieth-century glory with architectural aspirations to Roman grandeur, experienced rapid economic collapse. The abandoned heart of the city has become enjungled, and the area now annually attracts thousands of ruin-tourists, some of whom take 'urb-ex' (urban-exploration) tours through the city's dilapidating edifices. Upon Detroit in particular have descended hundreds of photographers, film-makers, artists, writers and other voyeuristic gleaners in the debris, attracted by the idea that Detroit offers a localized version of the world without us – a present-day future ruin – but often paying scant attention to the ongoing economic-social circumstances that frame the startling sights. So ubiquitous has the 'Detroit Jungle' become as an image, in fact, that – like Macaulay's New Zealander – it has become denounced as cultural cliché and worse (the sub-genre of 'ruin-porn' has been named and shamed as visually exploitative and ethically insensitive).

There is, unmistakably, a misanthropic comfort in such deep-green dreams of a post-human world. For when imagining our own deletion, we are absolved of our monstrous success as a species: forever out-expanding our niche and bringing about as we have the sixth great extinction pulse in the Earth's history. Our biotic hegemony can temporarily be forgotten, and the imagined return of nature offers temporary atonement. It will all be all right in the end, such fantasies assure us – the planet will recover from its infestation of *Homo sapiens.*

Various shades of green

Let me use that idea of hoped-for deep-green future to draw us out of the contemporary and back to 1885, thirteen years after Doré's engraving, and the year in which the English naturalist, journalist, essayist and novelist Richard Jefferies publishes his strangest book. *After London; or Wild England* is a novella in which, as result of an unspecified catastrophe, the English landscape has been dramatically re-wilded. 'The old men say their fathers told them that soon after the fields were left to themselves a change began to be visible', the book begins, in an opening whose tone is pitched partway between fairytale and lore, and which bears quoting at length (not least for its anticipation of this essay's preliminary counter-factual):

> It became green everywhere in the first spring, after London ended, so that all the country looked alike. The meadows were green, and so was the rising wheat which had been sown, but which neither had nor would receive any further care. Such arable fields as had not been sown, but where the last stubble had been ploughed up, were overrun with couch-grass, and where the short stubble had not been ploughed, the weeds hid it. So that there was no place which was not more or less green; the footpaths were the greenest of all, for such is the nature of grass where it has once been trodden on, and by-and-by, as the summer came on, the former roads were thinly covered with the grass that had spread out from the margin.
>
> In the autumn, as the meadows were not mown, the grass withered as it stood, falling this way and that, as the wind had blown it; the seeds dropped, and the bennets became a greyish-white, or, where the docks and sorrel were thick, a brownish-red. The wheat, after it had ripened, there being no one to reap it, also remained standing, and was eaten by clouds of sparrows, rooks, and pigeons, which flocked to it and were undisturbed, feasting at their pleasure. As the winter came on, the crops were beaten down by the storms, soaked with rain, and trodden upon by herds of animals.
>
> Next summer the prostrate straw of the preceding year was concealed by the young green wheat and barley that sprang up from the grain sown by dropping from the ears, and by quantities of docks, thistles, oxeye daisies, and similar plants . . . Footpaths were concealed by the second year, but roads could be traced, though as green as the sward, and were still the best for walking, because the tangled wheat and weeds, and, in the meadows, the long grass, caught the feet of those who tried to pass through . . .
>
> Aquatic grasses from the furrows and water-carriers extended in the meadows, and, with the rushes, helped to destroy or take the place of the

> former sweet herbage. Meanwhile, the brambles, which grew very fast, had pushed forward their prickly runners farther and farther from the hedges till they had now reached ten or fifteen yards. The briars had followed, and the hedges had widened to three or four times their first breadth, the fields being equally contracted. Starting from all sides at once, these brambles and briars in the course of about twenty years met in the centre of the largest fields . . .
>
> No fields, indeed, remained, for where the ground was dry, the thorns, briars, brambles, and saplings already mentioned filled the space, and these thickets and the young trees had converted most part of the country into an immense forest. Where the ground was naturally moist, and the drains had become choked with willow roots, which, when confined in tubes, grow into a mass like the brush of a fox, sedges and flags and rushes covered it. Thorn bushes were there, too, but not so tall; they were hung with lichen. Besides the flags and reeds, vast quantities of the tallest cow-parsnips or 'gicks' rose five or six feet high, and the willow herb with its stout stem, almost as woody as a shrub, filled every approach.
>
> By the thirtieth year there was not one single open place, the hills only excepted, where a man could walk, unless he followed the tracks of wild creatures or cut himself a path. The ditches, of course, had long since become full of leaves and dead branches, so that the water which should have run off down them stagnated, and presently spread out into the hollow places and by the corner of what had once been fields, forming marshes where the horsetails, flags, and sedges hid the water.
>
> (Jefferies 1885: 1–5)

Only a few humans have survived the scarification of their species. Mad-Maxish tribes of 'gypsies' and 'Bushmen' roam the land, divided along ethnic as well as self-interested lines. Jefferies' book pursues its lone adventuring hero, Sir Felix Aquila, who crosses a vast inland lake to reach at last a putrid and noxious swamp, which he realizes to be the site of 'the deserted and utterly extinct city of London', now lying 'under his feet' (Jefferies 1885: 378). The capital is granted no reprieve by Jefferies; Aquila is there to witness and thus authenticate its total vanquishing by nature. The pleasure with which Jefferies visits destruction upon human structures, and London in particular, is unmistakable. Various reasons suggest themselves for this, beyond his natural green-mindedness. He was, at the time of writing *After London*, suffering from tuberculosis

FIGURE 7.3 Edward Thomas, studio portrait by E. O. Hoppe, c. 1912. Reproduced with permission of Richard Emeny.

which he figured to himself, fascinatingly, as a 'dust' that settled on him (and which would kill him two years later in 1887). He held London partly responsible for that illness, as he also held the swiftly expanding capital responsible for the destruction of wildlife in its environs.

After London ran through numerous editions, and among those it influenced was a writer who is not normally thought of as a ruinist, Edward Thomas (1878–1917; Figure 7.3). Thomas remains best known as the author of such often-anthologized poems as 'Adlestrop' and 'As the Team's Head-Brass', and is popularly thought of as a pastoral poet-elegist for a rural England that was disappearing even as he wrote. He was, however, much more than this reputation suggests: he was a novelist, memoirist, short-storyist, biographer, travel writer and nature writer. His slim output of poetry is at last being understood as acutely modern in its forms and concerns; more interested in fragment than in integrity, and more in transit and displacement than in dwelling and nostalgia.

From a young age, Thomas was a naturalist; from a young age, he was a ruinist; and from a young age, he was a noticer. The three talents converge in a moment in his late essay on the seventeenth-century antiquary John Aubrey, which Thomas starts by praising Aubrey's talent for isolating telling details: 'who but Aubrey', Thomas asks, 'would have noticed and entered in a book the spring after the fire of London "all the ruins were overgrown with an herb or two, but especially with a yellow flower, *Ericolevis Neapolitana*?"' (Thomas 1917: 90). Who but Thomas would have noticed that Aubrey had noticed that overgrowing of ruins with the flame-yellow flower? It speaks of Thomas's persistent preoccupation with life in ruins. Reading through some of the thousands of pages of his published prose, one finds him again and again imaginatively drawn to the subject. Although he worked on Welsh monuments during his brief spell in the heritage industry, and although the crumbled castles of Swansea and elsewhere do feature occasionally in his prose, his ruins tend not to be the grand singular dilapidations beloved of Romanticism. Rather, he typically describes following a modest country lane to its end, and finding there a dilapidated barn or cottage, an abandoned farm, or a crumbling chalk pit.

Thomas wrote a biography of Jefferies, he read *After London*, and he returned often to the novella's vision of a re-wilded and largely dehumanized England. At times, Thomas's writing is tinged with a milder version of Jefferies' misanthropy. He explicitly returns to Jefferies' vision, for instance, in *The South Country* (1909):

> I like to think how easily Nature will absorb London as she absorbed the mastodon, setting her spiders to spin the winding-sheet and her worms to fill in the graves, and her grass to cover it pitifully up, adding flowers – as an unknown hand added them to the grave of Nero. I like to see the preliminaries of this toil where Nature tries her hand at mossing the factory roof, rusting the deserted railway metals, sowing grass over

> the deserted platforms and flowers of rose-bay on ruinous hearths and walls. It is a real satisfaction to see the long narrowing wedge of irises that runs alongside and between the rails of the South-Eastern and Chatham Railway almost into the heart of London.
>
> (Thomas 1909: 98–9)

The longing in Thomas's voice ('I like to think... I like to see... It is a real satisfaction') is audible: the perturbing pleasure which he, like Jefferies, takes in imagining the end of humanity and the supremacy of nature. Here, grass – as so often in future-ruinist literature – serves both as concealer and healer, greenly unwounding the damaged earth. Notice, too, the attention paid by Thomas to present-day metonyms of this future take-over – the miniature ruins that he spots on 'factory roofs' and in railway sidings. Such places offered him tiny scrying-glasses in which possible post-human futures of the Earth might be glimpsed.

Thomas is arguably at his most interesting and most modern, however, when he cleaves free of Jefferies' hatred of humanity, seeming to find it too easy a response. In an essay on chalk pits, Thomas explicitly acknowledged the 'misanthropy' that often accompanied the admiration of 'works of men that rapidly become works of Nature' (Thomas 2011: 513). Among his most intriguing ruin-texts is 'A Tale', a short poem written in March 1915:

> There once the walls
> Of the ruined cottage stood.
> The periwinkle crawls
> With flowers in its hair into the wood.
> In flowerless hours
> Never will the bank fail,
> With everlasting flowers
> On fragments of blue plates, to tell the tale.
>
> (Thomas 2008: 73)

What appears to be a simple poem about natural 'absorption' of a human structure, on closer inspection rapidly complicates. There is, most obviously, an uneasy relationship between the real flowers and the porcelain pattern. The blue periwinkle seems somehow to have emerged out of the shattered crockery – artifice leapt to life – and to be trying to escape the site of ruin, rather than staying to commemorate it: it 'crawls... into the wood'. 'Crawl' is itself a troubling verb: plants typically 'creep', rather than crawling, an action more quickly associated – especially given the date of composition – with a wounded human. No, nature in this poem serves neither as hopeful resurgent nor as assiduous archivist. The poem claims that the flower and the bank between them never fail 'to tell the tale', but of course no tale is fully told here. Who lived in the cottage, what their lives were like, why the cottage has fallen: all these aspects of the story go unrecorded and apparently unknown. Implicitly, also, there is a third layer to the twinned metaphor of periwinkle and flower-pattern, which is that the tropes of the poem constitute the third pattern, the third kind of memory. Certainly the poem is told in shards – consider the breaking up of the sentence that occupies the last stanza, for instance – and the reader is left to perform what partial reassembly is possible. Reading becomes an (incomplete) repair or reconstruction.

Compensatory noticings

Four months after writing 'A Tale' Edward Thomas signed up, though doubly acquitted of the obligation to enlist by being married and by being thirty-six years old. He joined the Artists Rifles and trained in England until, in December 1916, a call came round for volunteers to go out and serve with the batteries at the front. Again Thomas chose danger. He joined 244 Siege Battery, co-commanding 150 men, and on 29 January 1917 he and the troops boarded the *Mona Lisa*, bound for Le Havre and thence the frontline near Arras.

In the ten weeks he was at or near Arras, Thomas kept a war diary, dashing down in a Walker's back-loop pocket book a few hundred words for each day. These jottings provide a vivid record of his observations and imagination at the front. Among Thomas's preoccupations during those ten weeks was the relationship between life and ruin. By 6 February 1917, he was approaching the frontline proper. He had left behind the continuous and curvaceous forms of the North Downs in Kent, where he had been stationed for the final weeks of his training, and had reached a war-torn landscape of rupture and fragment – a realm of ruin. He billeted in 'half-ruined barns', he watched aerial 'shrapnel bursts', he passed a 'ruined pigeon house . . . and what was manor house', and he picked his way round shell holes on the Arras road (Thomas 1983: 160, 161). He was inhabiting a future ruin – a thought-experiment made atrociously real.

Thomas's work at the front was mostly as a watcher or noticer: his perilous job was to take up observation posts in ruined buildings – in one case in a part-collapsed factory chimney – and then use his field glasses to examine what he described as 'snowy broken land with posts and wires and dead trees', trying to work out the locations of the German guns or to track the fall of his own gunners' shells (Thomas 1983: 161). It was dangerous work; vulnerable to shells and to snipers, as well as to the collapse of the structures in which he was hiding. '4 shells nearly got me while I was coming and going', he notes calmly in his journal of a day spent up the chimney (170). Up at the front he was working in ruins. Behind the lines he was living in ruins. And during time off, he became a recreational ruin explorer, wandering through abandoned houses in Arras, and itemizing the objects he found inside them: 'an engraved 1850 portrait hanging on wall high up, without glass broken', '[a] crucifix, statuette, old chair', and on 'the top storey of high house', a 'ruined cloth armchair' with 'a garment laid across it after shell arrived' (164, 163). He began to see literally in and through the optic of ruins: 'How beautiful' – reads a journal entry from 29 March, that uncannily anticipates both Ballard and Hiorns – 'like a great crystal sparkling and spangling, the light reflected from some glass which is visible at certain places and times through a hole in cathedral wall, ruined cathedral' (173; Figure 7.4).

Thomas was not blithe about his material circumstances – writing of 'ghastly trees and ruins', and of living among 'rubble, rubbish, filth' – but he does not seem to have been drastically discomforted by his newly textured world (Thomas 1983: 166, 167). Much of his stability seems to have derived from his naturalist's habits of attention to the life that persisted even in the ruin of the trenchscape. On 13 February 1917 he noted the 'hare, partridges and wild duck in field S.E. of guns' (162). 'Black-headed buntings talk', he wrote on 14 February, 'rooks caw . . . grass rustl[es] on my helmet through trenches' (162). He watched kestrels hovering in pairs above the trenches, while above them wheeled 'four or five planes' (163). His attuned ornithologist's ear, superb at species identification, picked out a 'chaffinch say[ing] "Chink" in the chestnut', 'Partridges twanging in the fields', and '2 owls in garden at 6' (164–5). He heard '[l]innets and

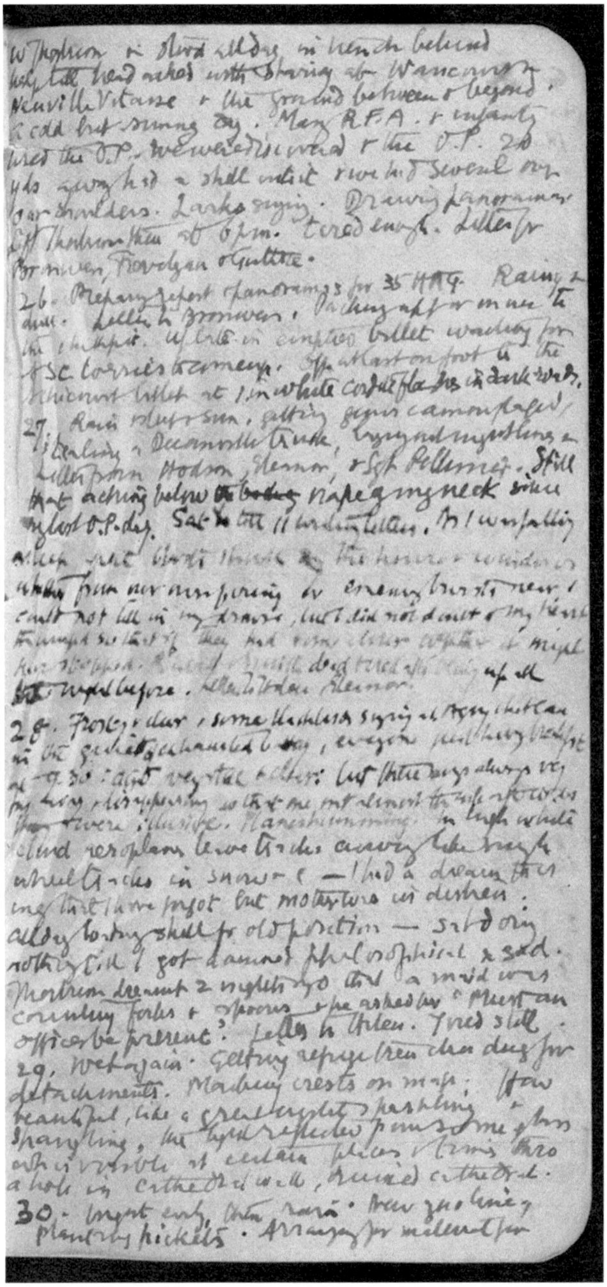

FIGURE 7.4 A page from Edward Thomas's war diary. The entry for '29' (29 March 1917) reads: 'How beautiful like a great crystal sparkling and spangling, the light reflected from some glass which is visible at certain places and times through a hole in cathedral wall, ruined cathedral.' The curved crease-marks visible on the page are thought to have been left by the air-blast from the shell that killed Thomas. This item is from the First World War Poetry Digital Archive, University of Oxford, www.oucs.ox.ac.uk/ww1lit, accessed 9 March 2014; © the Edward Thomas Literary Estate, and reproduced with thanks to the Digital Archive, and with permission of the Thomas Estate.

chaffinches sing in waste trenched ground . . . between us and Arras', and observed '[m]agpies over No Man's Land in pairs' (170). 'The shelling must have slaughtered many jackdaws but has made home for many more', he remarked on 23 February, a note that recalls a pre-war essay in which he imagined a ruined London as a 'pleasant' place in that it would provide dwelling places for jackdaws (165; Thomas 2011: 513). Reading the diary, one repeatedly encounters these compensatory noticings and narrations. The integrity of attention is used as a stay against the disintegrations of his context. Thomas's field-note-style diary jottings, broken in form, serve as a kind of ethical residue: fragments of concentration that might themselves, paradoxically, be whole and enough.

By late March 1917, the first major Arras offensive was approaching. Thomas had been at or near the front for six weeks, and ruin and nature had moved so close to one another in his imagination that they began to merge as perceptions. One of his last forward observation posts was in an old chalk pit. Another one was in a ruin, by a lilac tree, looking out through a hedge that provided his cover from snipers, and out of which blackbirds sang. When the German guns fired above his head and into Beaurains, he wrote, the shells came over 'like starlings returning, 20 or 30 a minute' (Thomas 1983: 171). A 'piece of burnt paper', which he saw blown towards him over no-man's land, turned out 'to be a bat shaken at last by shells from one of the last sheds in Ronville' (169). Machine-gun bullets 'snak[e] along – hissing like little wormy serpents' (174). Among the two things that Thomas recorded are that on 5 April the 'sods on f/c's dugout' had begun 'to be fledged with fine green feathers of yarrow' and then on 7 April that 'a great burst in red brick building in N.– Vitasse stood up like a birch tree or a fountain' (175). On the final pages of the diary, Thomas wrote single lines of verse, possible fragments of future poems. These were his last works, and as such they bear a family resemblance to the last work of another combatant-artist, the painter Rex Whistler. Whistler took part in the Normandy advance after the D-Day landings: shortly before he was killed, he pulled blue and red chalks from his battledress pocket and chalked a Madonna and Child on the exposed wall of a shelled church in northern France.

At dawn on 9 April 1917, Easter Monday, the first battle of Arras began. The morning was a triumph for Thomas's gunners. The British batteries disabled almost all of the German heavy guns with their counter-battery fire, and their troops took ground. As the guns fell silent the British soldiers emerged to shout and dance. Thomas stepped from his dugout, then leaned back into the doorway to fill and light his clay pipe. He had part-filled the pipe when a German shell fell near him and the vacuum caused by its passing threw him hard to the earth, killing him by pneumatic concussion. His body was left whole and unmarked, and beside it lay his clay pipe, unbroken. His war diary was recovered and returned to his family, and the folds visible on the pages of the diary are thought to be the pressure ridges caused by the fatal shell.

Modern ruinism and the ethics of futurology

Thomas might be taken to mark an important moment in the history of ruinism, in that he began to extricate himself both from the misanthropic absolutisms of apocalyptic ruinists such as Jefferies, and from the melancholic ruminations of Romantic ruinists such as Doré. Thomas was, or at least began to be, a modern ruinist: fascinated by the process and textures of ruin, but resistant to be being cast either nostalgically backwards or hatefully forwards. His writings on ruin, especially his war writings, are motivated by what Andreas Huyssen calls a 'remembrance of nature in all culture': by his realization that all our sites are, to some degree, sliding towards the organic, as Simmel proposed (Huyssen 2006: 13). We need, Thomas part-understood, to come to an accommodation with nature, rather than wishing for its utter eradication or total triumph, and in his writing he began – as Alan Weisman put it a century later – 'to dream of a way for nature to prosper that *doesn't* depend on our demise' (Weisman 2007: 5).

This essay opened with a thought-experiment, and it ends with consideration of another: Cormac McCarthy's 2006 post-apocalypse novel *The Road*, described by George Monbiot as 'the most important environmental book ever written' (Monbiot 2007). It imagines a future America brought to ruin by a swift pyrocaustic disaster of unspecified cause. So severe has been this calamity, however, that nature is not resurgent but obliterated. The landscape has been charred to cinders. No 'fine green yarrow feathers' fletch the land here. The light is grey, the forests are burnt, ash blows in loose swirls over the blacktop. The earth has been scorched back to its blackened residues, and so too has the novel's language, which exists as a rubble of paragraph blocks and verbless sentences. Through this ruinscape trek two refugees, a father and son, moving in order to keep out of the clutches of other survivors, and to find food. In the opening pages of the novel we watch the father keeping to cover, and using his binoculars to 'glass' the landscape ahead of him: 'Looking for anything of color. Any movement. Any trace of standing smoke' (McCarthy 2006: 4). He is a version of Edward Thomas, ninety years on: the watcher scanning no-man's land, looking for life amid the ruins.

Only three sources of hope survive in McCarthy's bleak novel. The first is narrative itself, hopeful in terms of the persistent paradox of apocalyptic art, which is that to annihilate the world artistically one must simultaneously summon it into being. The second source of hope is the boy, who the father keeps alive and who keeps the father alive, and promises the possibility of a future after ruin. Third are the father's memories of a better and earlier world, a world before the fire, a world not wholly unlike our own, in which human relations with nature have not yet been irrevocably broken. In the closing paragraph of the novel, after a desperate journey through the ashy land towards a dark future, the narrative voice – its source shifting and unsure – casts back to that earlier world – notionally our present world – and recalls, for the second time in the novel, 'brook trout' finning against the current in a stream:

> They smelled of moss in your hand. Polished and muscular and torsional. On their backs were vermiculate patterns that were maps of the world in its becoming. Maps and mazes. Of a thing which could not be put back. Not be made right again. In the deep glens where they lived all things were older than man and they hummed of mystery.
>
> (McCarthy 2006: 241)

The language itself here hums with mystery – moss, muscular, maps, mazes, man, becoming – and is itself syntactically mysterious. Read through, the 'thing which could not be put back' is at first the fish themselves, there in 'your hand', lifted from the stream – but it is also 'the world in its becoming': 'put back' in the sense of 'made right'. The prose serves here as elegy and celebration, warning and plea, tocsin against toxin: do not lose what you have before it is too late for it to be repaired or to be restored. 'We run the danger of bearing witness to history too late to effect change', as Jean Baudrillard observed (Baudrillard 1989: 33). McCarthy's novel – like other constructive visions of life in ruins – might enable us to bear witness to our own future history, and in this way help us also to avoid ruining life.

References

Anon. 1865. 'A Proclamation', *Punch* 48 (7 January 1865): 9.

Ballard, J. G. 2008. *The Drowned World.* London: Flamingo. Orig. pub. 1962.

Baudrillard, J. 1989. 'The Anorexic Ruins', in D. Kampfner and C. Wulf, eds, *Looking Back on the End of the World,* trans. D. Antal. New York: Semiotext(e). 29–45.

Dillon, B. 2011. 'A Short History of Decay', in B. Dillon, ed., *Ruins.* London: Whitechapel Gallery. 10–19.

Huyssen, A. 2006. 'Nostalgia for Ruins', *Grey Room* 23 (2006): 6–21.

Jefferies, R. 1885. *After London; or, Wild England.* London: Cassell.

Macaulay, T. B. 1840. 'Leopold von Ranke's *The Ecclesiastical and Political History of the Popes During the Sixteenth and Seventeenth Centuries* [*Die römische Papste*] trans. S. Austin', *Edinburgh Review* 72 (1840): 227–58.

McCarthy, C. 2006. *The Road.* London: Picador.

Monbiot, G. 2007. 'Civilisation Ends with a Shutdown of Human Concern. Are We There Already?' Comment is free, *The Guardian,* 30 October 2007. www.guardian.co.uk/commentisfree/2007/oct/30/comment.books. Accessed 9 March 2014.

Thomas, E. 1909. *The South Country.* London: J. M. Dent.

1917. *A Literary Pilgrim in England.* New York: Dodd, Mead and Company.

1983. 'Diary 1 January – 8 April 1917', in R. G. Thomas, ed., *The Childhood of Edward Thomas.* London: Faber and Faber.

2008. *The Annotated Collected Poems.* Ed. E. Longley. Tarset: Bloodaxe.

2011. 'Chalk Pits', in G. Cuthbertson and L. Newlyn, eds, *Edward Thomas: Prose Writings (A Selected Edition),* vol. 2: *England and Wales.* Oxford: Oxford University Press. 512–18.

Weisman, A. 2007. *The World without Us.* London: Virgin.

Woodward, C. 2001. *In Ruins.* London: Chatto and Windus.
2011. 'Learning from Detroit, or The Wrong Kind of Ruins', in A. Jorgensen and R. Keenan, eds, *Urban Wildscapes.* London: Routledge. 17–33.

Further reading

Doré, G., and B. Jerrold. 1872. *London: A Pilgrimage.* London: Houndsditch.
Mellor, L. 2011. *Reading the Ruins.* Cambridge: Cambridge University Press.
Simmel, G. 1965 (1911). 'The Ruin' (*Die Ruine: Ein ästhetischer Versuch*), in Georg Simmel, *Essays on Sociology, Philosophy and Aesthetics.* Ed. K. Wulff, trans. D. Kettler. New York: Harper and Row. 256–66.
Skilton, D. 2004. 'Contemplating the Ruins of London: Macaulay's "New Zealander" and Others', *Literary London Journal* 2 (1). http://homepages.gold.ac.uk/london-journal/march2004/skilton.html. Accessed 9 March 2014.

8 The after-life*

CLIVE GAMBLE

Be prepared

When asked about what happens next, Woody Allen replied, 'I don't believe in an after-life, although I am bringing a change of underwear.' Such preparedness goes to the heart of understanding attitudes to the after-life. 'Leave nothing to chance' was the advice taken by the ancient Egyptians who could afford a Book of the Dead to be buried with them, even if the book factory where they bought it often misspelt their names (Taylor 2010). The words, it seems, did not matter as much as the possession of the object. As a satnav to the future such devices could never be properly tested, or returned as faulty, and many purchasers probably got stuck on the equivalent of a bridge that was too narrow in the netherworld of the dead. Nonetheless, the book which contained some 200 spells had a long shelf life, enjoying a period of popularity between 1550 and 1069 BC. It acted to overcome the dangers of the netherworld, which were many, and through which the dead person walked, floated on air or travelled by boat. Clearly the inspiration for today's gaming industry, the many levels and dangers that the dead person encountered were overcome by the special powers of the spells that transformed them into a variety of animals and plants; and by transforming they triumphed until they reached their goal and the game was over.

After-life and after-person

As an archaeologist I could provide an historical account of the many and varied beliefs in the after-life. I could range across the cultures of the

* The paper benefited from discussions with Lyn Ellett, Gustaaf Houtmann and Elaine Morris, and financial support from the British Academy Centenary Project From Lucy to Language the Archaeology of the Social Brain and my Co-Directors Robin Dunbar and John Gowlett.

world, digging into the temporal archives to recover nuggets of information that would delight and intrigue, and which gathered together would form a facet of its creator's after-life; read, transformed and remembered.

But I am not going to take that road. Instead, I want to approach the after-life as a geographer with an interest in the spaces and people of the past; as Bob Dylan, using updated Egyptian references, puts it in *Blonde on Blonde* (1966):

> Oh, Mama, can this really be the end
> To be stuck inside of Mobile
> With the Memphis blues again.

From a geographical perspective the after-life is a landscape that has cognitive, social and aesthetic properties comparable to any in the present-life. I will argue that the after-life is an *imaginary geography* no different in principle to Alice's Wonderland or James Cameron's Pandora.

Edward Said (1978: 55), who famously dissected Orientalism with the concept, maintained 'there is no doubt that imaginative geography and history help the mind to intensify its own sense of itself by dramatizing the distance and difference between what is close to it and what is far away'. Said's observation points us in the direction I will take. The after-life is a creative exercise and one that stems from a basic dramatic property of the mind and the spatial mechanisms by which we construct an identity, a sense of self. However, I will argue for a different understanding of the mind than one which allows it 'to intensify its own sense of itself'. The mind, as something inside our heads, intensifies nothing especially itself. On the contrary, and according to George Lakoff and Mark Johnson (1999: 3), cognitive science has demonstrated the following;

> (1) that the mind is inherently embodied, thought is mostly unconscious, and
> (2) abstract concepts are largely metaphorical.

As an archaeologist, I would go further still and introduce the concept of the after-person. If the after-life is an exercise in imaginary geography then the concept of the after-person is about the agency of things. This distinction, as I will show later, is necessary if we are to understand the deep history of belief as revealed by archaeological evidence.

The capacity to deal with both depends on conceiving the mind as an embodied, extended artefact, distributed widely through materials, objects and other identities rather than simply contained in the skull or networked in neurons. It is both rational and relational, which means that it depends on context and emotional engagement to make sense of the information it receives, the associations it fashions and the geographies it peoples. An extended mind is an inclusive project combining the materials encountered in the world, the body's chemically based emotions and its interior thoughtscapes. It is these combinations that achieve Said's amplification and the outcome has indeed been to dramatize the distant and the strange by conjuring our imagination.

Therefore, in order to understand the landscape of the after-life and the agency of the after-person we need to investigate why humans have such a capacity for imagination. Why is it that we use metaphor and symbol to structure our lives, including those in the past and the one to come, in an imaginative rather than strictly empirical way? Moreover, I do not believe that during our evolution over 6 million years we always had this exaggerated capacity. How and when it evolved is of interest, because of the underlying principles, rather than the pursuit of after-life beliefs for their own sake. Consequently, I will argue that human imagination emerged as a property of social life long before there were Books of the Dead, visions of heaven and hell and the Memphis Blues. Here I will focus on the deep history of humans, when the heavy evolutionary lifting was done, and where the realms of the cognitive and social became entangled with those of the material and aesthetic. I will also focus on a time 600,000 years ago as the tipping point in hominin–human evolution for these realms.

Going beyond and hominin imagination

The hominin[1] capacity to think outside the box, to go beyond the immediate world where interaction is mediated by the senses, is one of our defining capacities. We are not the only animal either to think forward or

[1] Hominin includes ourselves (humans) and all our fossil ancestors such as the Neanderthals and Australopithecines. It excludes the great apes who are classed as hominids.

to draw on past experience. But we are the only animal to take it to such social extremes. This is the concept of *going beyond* the here and now to imagine either the activities of others who though alive are not present before us, or an ancestor who lives in our memories, among us but not with us. To 'go beyond' implies a distributed cognition that extends us across time and space. Such cognition involves not just living equivalents but the material networks in which we are enmeshed and without which we would be very different social beings.

This has long been recognized. In 1759 Adam Smith argued in *The Theory of Moral Sentiments* that our imagination is the key to our capacity for sympathy (I.I.2). In what today might be called a theory of mind, Smith declared that 'we have no immediate experience of what other men feel . . . By the imagination we place ourselves in his situation.' Materials and the senses, which are the core resources of the human social experience, are central to that human capacity.

This imaginative capacity for extension has been identified as central to understanding human social life by disciplines that have different approaches to the issue. Tony Giddens defines one of the principal questions for social theory as follows, 'how the limitations of individual "presence" are transcended by the "stretching" of social relations across time and space' (Giddens 1984: 40). The same issue emerges for primatologists and where going beyond the information available to construct society is a feature of human social cognition (Quiatt and Reynolds 1993). In a comparison of human and primate communities a team of behavioural ecologists recognizes the importance in the social life of the former of uncoupling relationships from the constraints of co-presence; what they term the release from proximity (Gamble 1998; Rodseth et al. 1991).

Our primate heritage requires that we construct social bonds on the basis of personal interaction and where their strength derives from the emotional legacy involved in such encounters. But notoriously we go beyond this flesh-and-blood intimacy to stretch these bonds when partners are absent and to confer equal status to surrogates. This sense was captured for the Facebook generation by the Greek poet Palladas who lived in Alexandria during the fourth century AD:

> Loving the rituals that keep men close,
> Nature created means for friends apart:
> Pen, paper, ink, the alphabet,
> Signs for the distant and disconsolate heart.[2]

As a result we can maintain equally intense relationships with artefacts such as sports cars, smartphones and family pets. A recent study, widely reported in the British press in August 2011, claims that 18 per cent of the women surveyed spent more time talking to their dog than their husband or boyfriend.

The achievement of hominins has been two-fold. First, we have been able to amplify the bonds of social life so that they have power to condition interaction *in absentia*. Second, we have enfolded into our social universes the agency of inanimate materials and non-human organisms. Long derided as symptomatic of the primitive mind the practice of animism, magic, sorcery, taboo and totem enshrine an understanding, suppressed in our own scientific culture, of the power of things. Such agency, as anthropologist Alfred Gell (1998: 96) pointed out, is a different form of agency to that between two hominin actors. But so long as objects, animate or inanimate, are implicated in networks of social relations then they are both the source *of* and target *for* agency. And if agency is an expression of our capacity for social life then this is scaffolded around the core of materials and senses; aesthetics and emotions; the battery, if you like, that powers the clockwork rabbit of human imagination.

Symbols and metaphors

Understanding the hominin capacity for imagination that transcends quotidian life and peoples fabulous landscapes such as the ancient Egyptian netherworld and Michael Jackson's Neverland, is therefore as much a material as a mental process. It depends on thoughts and things and the relationships drawn between them in the many and varied social contexts. These landscapes are both symbolic and metaphoric; rich in signs

[2] Translated by Tony Harrison and appearing as a 'Poem on the London Underground' in 2011.

and substitutes but also grounded in sensory experience. When seeking the appearance of hominin imagination archaeologists have stressed the discovery of objects as signs. These are highly varied ranging from some tracery on animal bones that seem more than the activity of de-fleshing to more obvious symbols of identity such as adornment, gifts and burials.

The paucity and profusion of such objects in different regions and at different times in the Palaeolithic world sparks controversy over the location, antiquity and makers of these objects. The greatest clash is between two camps; only-humans had symbols *and* other-hominins-than-humans had symbols. The aim is to demonstrate when and where the potential for so-called behavioural modernity arose.

I confess that I have grown rather weary of this debate and the endless claims and counter-claims (for a recent overview see Shea 2011). Neither do I understand what 'behavioural modernity' means. Can symbols only exist if they are made material? Could there ever have been a time when hominins had no symbols? Colin Renfrew (2007) has suggested there was a time in the Palaeolithic when there were symbols but no modernity. Did that come later? Perhaps in the Neolithic when people were now densely packed in villages and towns farming the fields around. Urbanism concentrated minds wonderfully. The result was an explosion of human imagination and imaginary geographies.

The problem is that signs and symbols can be readily identified linguistically but remain opaque when it comes to objects. We are left trying to determine the symbolic intentions of someone, 400,000 years ago, holding a bone in one hand and a flint tool in the other. What we are being asked to accept or reject is a rational account where the author of the scratches deliberately chose between a symbolic and a non-symbolic gesture; to depict or to de-flesh. But can such a decision ever be made? How could one act (depiction) be more symbolic than another (de-fleshing a bone)? That type of enquiry is beyond archaeology; not because the evidence cannot survive but because it supposes a universal cognitive process identical to our own. That process makes familiar distinctions between categories such as sacred and profane, symbolic and mundane, person and thing. Cognition is separated from the world rather than extended into it.

An alternative is to switch the vocabulary away from symbol and towards metaphor. The basis of metaphor is that we *experience* one thing in terms of another (Gallese and Lakoff 2005; Lakoff and Johnson 1980; Lakoff and Johnson 1999). As a result, the embodiment of our senses becomes important in the way metaphors derive their meaning. Instead of a semiotic approach, where signs and symbols are interrogated for their significance, the opportunity presented by metaphor is to return the senses to the study of relational knowledge. Furthermore, while semiotics is essentially a study based on our capacity for language and then applied to objects, a metaphorical approach draws on linguistic, material and musical sources. These three sources for metaphor can be highly nuanced and entangled.

Let me give you an example. Many metaphors arise from our experience of being contained, enfolded: for example, 'I'm in a jam', 'To be stuck inside of Mobile', or Winston Churchill's 'Russia is a riddle, wrapped in a mystery, inside an enigma'. Many material metaphors are also containers: the tattoos on our skin, the clothes we wear, the drums we beat, the houses we live in, the presents we wrap to exchange. These solid metaphors dominate contemporary culture. They have done since the container explosion of pots, bags and baskets that appeared with the first farming, displacing the instrument, such as a spear or knife, as the favoured material metaphor. This dominant way of thinking through objects has also permeated concepts. How could there be kinship categories without a notion of containment (Gamble 2008), or the institutions of the State that enfold the citizen?

Objects before words

Therefore, the re-alignment from a focus on symbols to an interest in metaphors has immediate pay-offs for considering the evolution of hominin imagination that pre-requisite for an after-life. Archaeologists have a clear answer to the question: which came first, linguistic or material metaphors? Things, solid metaphors, preceded words and linguistic symbols. As recently as 1.5 million years ago hominins, such as the Nariokotome boy from West Turkana in Kenya, had developed bipedal walking and the capability for endurance running (Bramble and Lieberman 2004).

These *Homo erectus* hominins also made symmetrical stone implements, material metaphors that took the form of instruments, extensions of the limb. However, a study of his hypoglossal canal preserved in his vertebral column shows he lacked fine muscle control over his tongue (MacLarnon and Hewitt 2004). Consequently he was unable to control his breathing to produce the significant pauses in speech production that contribute to language. That came later, and certainly by 600,000 years ago when large-brained hominins were widespread across the settled portions of the Old World.[3]

The size of the brain is more important than the detail of the faces of these early hominins. As shown by Aiello and Dunbar (1993) for apes and monkeys, the larger the brain, the larger the personal network sizes of an individual in those primate communities. The upper limit for primates is about seventy such interaction partners and reflects the conflicting calls on their time budgets – eating, socializing, travelling and resting. Using the same equation the prediction for small-brained hominins such as Nariokotome is 104 interaction partners and among the large-brained *Homo heidelbergensis*, ancestor to ourselves in Africa and Neanderthals in Europe, it rises to 126. *Homo sapiens* with larger brains again has a personal network size of 150, now known as Dunbar's number and confirmed by the average social-network size for Facebook users (Roberts 2010).

This cross-species comparison forms the basis of the social brain hypothesis (Dunbar, Gamble and Gowlett 2010) which proposes that our social lives drove encephalization and where a selective advantage came from reduced predator pressure (Dunbar and Shultz 2007). The result is that we have a brain some three times larger than expected for a primate of our body size. One consequence of larger personal network sizes was that the traditional forms of creating and maintaining bonds by finger-tip grooming became too time consuming. Therefore, by 600,000 years ago language had evolved to facilitate interaction among larger numbers of partners while at the same time overcoming an important time-constraint (Aiello and Dunbar 1993).

[3] Nariokotome is a juvenile *Homo erectus* with a brain size of 909 cm^3. By 600,000 years ago the large-bodied *Homo heidelbergensis* had a brain size of at least 1200 cm^3.

The four FACEs of the extended mind

However, in this brief excursion into the antiquity of language we need to remember that long before words there were objects. The earliest stone tools are currently 2.5 million years old (de la Torre 2011; Semaw et al. 1997) and could extend back to 3.3 million as cut marks on animal bones hint from the Ethiopian locale of Dikika (McPherron et al. 2010). However, such simple cores and flakes are never going to make it into an archaeologist's lexicon of symbols. Much later the Nariokotome boy, 1.5 million years ago, was contemporary with developments in stone-working that saw basalt outcrops and river cobbles knapped to produce large cutting tools, some of which display marked symmetry. This has led some to argue for a symbolic dimension (Carbonell et al. 2003) but the majority prefer a functional explanation such as better butchers' knives (White 1998) or indicators of reproductive fitness (Kohn and Mithen 1999).

However, very similar objects were being made almost a million years later by *Homo heidelbergensis* at that critical time of 600,000 years ago. The novelty by this time was the maintenance and containment of fire, a development that added time to the social day (Gowlett 2010), and the common use of tools composed of different materials (stone, wood and bindings) to produce projectiles (Barham 2010). Such composite artefacts, like an after-life, do not occur in nature. They exist because technology is embodied, as Marcel Mauss showed a century ago, and so grounded in the mundane practices of social life. Those simple stone-tipped spears are evidence of a developed hominin imagination that can fashion new identities by making associations. As we relate so we create.

Palaeolithic burials provide another example (Pettitt 2011). Several Neanderthal skeletons, dating to between 60,000 and 100,000 years ago, mimic Christian burial practice; a dug pit, a complete body and possibly some grave goods in the form of flowers and food. Prior to these burials the evidence has a degree of ambiguity; bodies accumulated at the bottom of a shaft and bits and pieces. Paul Pettitt describes these data as an archaic mortuary phase, possibly as old as 600,000 years, and typified by a *Cronos* compulsion expressed in de-fleshing and sometimes eating carcasses. Such embodiment, undertaken for a number of reasons but principally out of an interest in morbidity (Pettitt 2011: 8–9), speaks to the concept of the after-person rather than the after-life. This concern with the causes and nature of someone's death implies an agency that does not depend on life but on relationships created between people and material stuff.

It is through close emotional engagement with stuff, materials and the artefacts fashioned from them, that we construct our social lives. These imaginative identities take the form of four interlinked social practices, the FACE of the extended mind: **F**ragmentation, **A**ccumulation, **C**onsumption and **E**nchainment. The point about FACE and the extended mind lies in the metaphorical, experiential basis of these practices.

These four FACEs explain why archaeological evidence has such robust, familiar and repeated patterns that take the form of well-structured sets and nets of artefacts. Viewing artefacts as material metaphors provides a common understanding between past and present based on embodied experience. To make such artefacts stone has to be fragmented. And once fragmented there is the opportunity to enchain, to create social relationships through simultaneously building sets and nets of artefacts with others, across landscapes and deliberately at places. The metaphorical basis of a simple action such as fragmentation is, as John Chapman and Biserka Gaydarska (2010) argue, that of metonymy and synecdoche, where parts stand for wholes. To break open an animal carcass to get at the food inside is to fragment the whole. To pass the meat and marrow around the hunters and their families is to enchain (Gamble and Gaudzinski 2005). Enchainment and fragmentation simultaneously created sets and nets of material of bones and stones in time and across space. They were augmented by the other relational practices that involve people and things: accumulation and consumption (Gamble 2007). These are evident at the earliest hominin locales where materials are little disturbed by erosion and movement, some of them are even *in situ* after the lapse of a million years. And although these archaeological sites are very different from a hunting camp today in the Kalahari or northern Canada, the underlying principles are the same.

As a final example of this imaginative FACEworld I will give you the example of grog. Not alcohol, but ground up pottery shards for use as temper to add to the clay to assist the firing of pots. Grog is the only temper that does not occur in nature. It is incorporated into the clay matrix, a fragment of an existing pot, accumulated as part of a new one. Grog also enchains because the ground-up pot belonged to a dead person. Indeed the death of a potter often results in her pots being broken and the pieces passed around (Sterner 1989). In this way the generations are reproduced in the manufacture of new pots contained in a cycle of production and re-production. And as consumption proceeds, because the pot is still a container for liquids and food, the generations are carried forward (Morris 1994); an after-life squeezed into a surrounding clay wall. And when the surface of the pot is decorated then the ancestor within is protected, wrapped in another layer of FACEworld.

After-life, after-person and theory of mind

For sociologists and anthropologists 2012 marked the centenary of Emile Durkheim's *The Elementary Forms of the Religious Life*. The central message in his global review was that religious life is not a special, separate activity. It marks neither a transition in human development (for example, from primitive animism to civilized religion) nor requires its own vocabulary and concepts to achieve understanding. Instead religious behaviour, in all its varied ethnographic forms, lies firmly rooted in the practices of social life. To analyse the social is at the same time to understand the religious dimension. Religion is an example of the buzz of the social gathering, what Durkheim called *effervescence*, and which provides both an explanation for the assembly and the power of association that lingers long after the participants have dispersed.

Viewed from this perspective religious life is an example of amplification, as discussed above. At the core of social life lie the bonds between dyads, such as mother and child, and these bind due to the material and sensory resources available; for example, food, touch and the sense of well-being. Closely associated with this core are the aesthetic qualities of materials and the emotions ascribed to social context. It is around this core that social life and all other imagined lives are scaffolded. These resources are then available for amplification into the social forms we are familiar with such as kinship, icons, laughter and music, the bricks in a building and origins myths.

These basics, the common core, have remained constant throughout hominin evolution. But social cognition has changed as any comparison between primates and humans shows. I would single out one aspect as important, a theory of mind, since this human capacity helps us understand our ability to imagine after-lives.

Theory of mind is, as Adam Smith pre-figured, an imaginative exercise. It depends on recognizing that another person has a different perspective than you do; in other words they have a mind. This cognitive capacity appears in children between the ages of three and four as manifest in the ability for false belief. Before this there is self-awareness, a capacity shared with several other social animals including the great apes. Whether great apes such as chimpanzees have false belief and therefore a theory of mind is much debated as is the importance of language for human mentalizing. The majority view is that chimpanzees don't, and that language is critical.

Language is how we understand advanced theory of mind with higher levels of social reasoning. For example, the statement 'Peter *knew* that Sarah *wanted* to put flowers on Norman's grave because God *desired* it' has three levels of intentionality. It shows complex social reasoning and the capacity to attribute mental states to others and goes beyond the child of five who is capable of recognizing false beliefs.

However, as cognitive studies of human development show we do not have to wait long for a belief in the agency of after-persons. Jesse Bering and Becky Parker (Bering 2006; Bering and Parker 2006) tested three age groups of children (3–4-, 5–6- and 7–9-year-olds) to see when they reach the cognitive capacity for symbolic causal reasoning; in other words the ability to recognize an external agency. In their study they told children that a fairytale person, Princess Alice (it could have been a dead relative but for obvious reasons was not), would give them a sign if they made the wrong choice about which of two boxes contained a ball. The sign was either a flashing light or Princess Alice's picture falling off the wall once the child's hand started to move towards the wrong box. The measure was to see if they then changed their choice. Only seven-year-old children understood this sign as symbolic of an intention by Princess Alice to communicate. The four-year-olds, for example, just thought the picture wasn't properly stuck to the wall. Bering and Parker (2006: 259) conclude that 'it is not the capacity to detect agency behind unexpected events that is late developing but the capacity to see communicative meaning in unexpected events'. In short that imaginative leap which attributes mental states to others whether they are in this-life or an after-life, real or virtual, animate or inanimate, living people or after-persons. And not only ascribing mental states but allowing them the power to affect the lives of the living.

But level-3 intentionality is only the start. Upwards lie the mental gymnastics where humans exhibit their relational skills as shown in complex plotlines such as Raymond Chandler's *The Big Sleep* and the realm of myth (Knight 1983). Such levels of intentionality are truly mind-bending but commonplace.

But which hominin ancestors had theory of mind and was it an advanced level? Did they all have a belief in the after-life and the after-person? If language was essential then, as I have already suggested, social reasoning akin to theory of mind could be at least 600,000 years old. The problem is that if they did they are not letting on. The artefact record does not change significantly at this time as we might expect, a bit like Princess Alice's falling picture, to indicate that advanced levels of intentionality were now present.

However, using the social-brain model Dunbar has provocatively suggested that level 2, recognizing another mind, occurs in hominins with brains larger than 500 cm^3; a value not achieved by chimpanzees. Level 3 he associates with brains over 1000 cm^3 and level 4 and above over 1400 cm^3. The critical link is between larger group size requiring novel forms of interaction such as language, and between language and higher order intentionality.

But what if materials and artefacts, the archaeologist's bread and butter, are not reliable signs to such a fundamental change? What if the other element in the social core, emotion, is a better guide?

Emotions do not preserve. Moreover they are slippery things to define. When does anger pass into rage or satisfaction into contentment? However, they can be usefully classed into three large families. At the base are mood emotions that are felt at places and about things and people.

Then there are the primary emotions which all animals have for good survival reasons: fear, happiness, disgust, anger. Finally, we come to the social emotions which depend on a theory of mind, recognizing another's point of view. These include guilt, shame and pride. Some animals are thought to have a sense of guilt. Usually the family dog after it has eaten the family's dinner. But that is something else. Would the wolf feel guilt if it ate Little Red Riding Hood in the forest? The family dog shows us that materials can be trained to feel as another facet of domestication. A dog by virtue of being taught emotional responses is an amplified material resource, which perhaps explains why many women prefer talking to them rather than their partner.

So 600,000 years ago artefacts did not need to change to record that theory of mind had appeared but the emotional basis of social life almost certainly did (Gamble, Gowlett and Dunbar 2011). Social reasoning of at least three orders of intentionality probably existed and moral codes, in the form of social emotions, now structured hominin society.

After-life and after-persons

Where does this leave the evolutionary changes in deep history? As we saw with the Egyptian Book of the Dead, the after-life is a personal, imagined journey into the unknown. Others will take the same road but the private, autobiographical aspect is what matters in this imaginary geography. By contrast the after-person is about the continuity of social cognition across the same boundary. Here it is not so much about what will happen to us but what happened to those who preceded us. Maintaining the flow of agency is the main concern and this is achieved by channelling it through the biography of objects and where people are a special case of this wider class of stuff. Both forms are nowadays commonplace, but from a hominin perspective of 600,000 years this might not have been the case.

First, the after-person. Bering (2006) argues that a belief in the continuity of mental states is the default position and that it is counter-intuitive to deny it. He bases his conclusion on a study of responses from six belief groups[4] who were questioned about the ability of aspects of consciousness to survive death. What is interesting from this psychological study is that even among the most hardline after-person sceptics (the extinctivists and agnostics) Bering found some support that the mental states of emotion, desire and knowledge continue after death. On the contrary the sceptics find common ground with the after-person believers (immortalists and reincarnationists) that the biological, psychological and perceptual characteristics of a person do not persist across the frontier.

What then is the evolutionary advantage of letting our imaginations run riot with such higher-order intentionality applied to after-persons? The question becomes one of adaptation. Is the after-person an unintended consequence, an exaptation, of advanced theory of mind? Or are they an adaptive mechanism serving a wider evolutionary goal? One suggestion is that such continuity was a mechanism for hyper-vigilance (that rustle in the bushes is more likely the sign of a leopard than the benevolent breeze). Or is it because we cannot cope with the cognitive void implied by the elimination of mental states? Is it what Pettitt (2011) refers to as a *Cronos* compulsion and which has a deep hominin history? We simply cannot imagine being without emotion, desire and knowledge so that in our memories the dead must also have these attributes. From there it is a short mental step to believing that supernatural beings, the top dogs in the after-life, can send signs of displeasure if the living break the social contract. These signs take the form of famine, flood and earthquake, or whatever events apply to your locality or culture. In short, 600,000 years ago after-persons were the moral guardians, the netherworld police, for those bigger social groups indicated by larger hominin brain sizes. And since the mechanism to amplify the bonds that bound these people together was weighted towards the emotional rather than material side of the social core, we should not be surprised that social emotions, underpinned by advanced theory of mind, united living and after-persons.

Second, the after-life. If after-persons are about continuity, albeit a bit disembodied, then the after-life is a response to going beyond. For

[4] These groups are described as non-believers (extinctivists, agnostics) and believers (immortalists and reincarnationists) as well as some eclectics and fence-sitters (consciousness survives death but uncertain what happens afterwards).

example, the ability to materially represent an after-life and an after-person are present in the grog-tempered pot. But we can do better than such recent examples. Without the ability for social extension, made possible by our distributed cognition and our means of amplification through materials, via FACE, and under evolutionary pressure, we would be a very different species. Most notably our expansion from an Old World resident for 6 million years of our evolution to becoming a global settler in only the last 50,000 years has been as dramatic as the rise of cities, numeracy, literacy and civilization. Worldwide humans came long before the worldwide web.

This capacity to expand and become a global species is one of *Homo sapiens*'s greatest achievements. Other developments were needed: the technology of boats, the flexible container of kinship, the ability to accumulate and store food and so defeat seasonality. But there was also, I argue, an imaginary geography of somewhere to go as well as the continuity of invisible action. The after-life is a parallel projection of that global itinerary; a journey accomplished by mobile peoples, who lived at low population densities, fishing, gathering and hunting but equipped with the means to think ahead; materially through the metaphors of containers and instruments, linguistically through signs and symbols and emotionally able to react to the challenge. The after-life, but not the after-person, began with those first footsteps 50,000 years ago beyond the boundaries of the hominin world and into the virtual and real worlds of humans, those imaginary geographies. And why? Because as Jack Kerouac put it in *On the Road* (1957: 25), 'there was nowhere to go but everywhere'.

Further reading

Aiello, L., and R. Dunbar. 1993. 'Neocortex Size, Group Size and the Evolution of Language', *Current Anthropology* 34: 184–93.

Barham, L. S. 2010. 'A Technological Fix for "Dunbar's Dilemma"?', in R. Dunbar, C. Gamble and J. A. J. Gowlett, eds, *Social Brains and Distributed Mind*. Oxford: Oxford University Press. 367–89.

Bering, J. M. 2006. 'The Cognitive Psychology of Belief in the Supernatural', *American Scientist* 94: 142–9.

Bering, J. M., and B. D. Parker. 2006. 'Children's Attributions of Intentions to an Invisible Agent', *Developmental Psychology* 42: 253–62.

Bramble, D. M., and D. E. Lieberman. 2004. 'Endurance Running and the Evolution of *Homo*', *Nature* 432: 345–52.

Carbonell, E., M. Mosquera, A. Ollé, X. P. Rodríguez, R. Sala, J. M. Vergès, J. L. Arsuaga and J. M. Bermúdez de Castro. 2003. 'Les premiers comportments funéraires auraient-ils pris place à Atapuerca, il y a 350,000 ans?', *L'Anthropologie* 107: 1–14.

Chapman, J., and B. Gaydarska. 2010. 'Fragmenting Hominins and the Presencing of Early Palaeolithic Social Worlds', in R. Dunbar, C. Gamble and J. A. J. Gowlett, eds, *Social Brain and Distributed Mind.* Oxford: Oxford University Press. 413–47.

de la Torre, I. 2011. 'The Origins of Stone Tool Technology in Africa: A Historical Perspective', *Philosophical Transactions of the Royal Society of London B* 366: 1028–37.

Dunbar, R., C. Gamble and J. A. J. Gowlett. 2010. 'The Social Brain and the Distributed Mind', in R. Dunbar, C. Gamble, and J. A. J. Gowlett, eds, *Social Brain and Distributed Mind.* Oxford: Oxford University Press. 3–15.

Dunbar, R., and S. Shultz. 2007. 'Evolution in the Social Brain', *Science* 317: 1344–7.

Gallese, V., and G. Lakoff. 2005. 'The Brain's Concepts: The Role of the Sensory-Motor System in Conceptual Knowledge', *Cognitive Neuropsychology* 22: 455–79.

Gamble, C. S. 1998. 'Palaeolithic Society and the Release from Proximity: A Network Approach to Intimate Relations', *World Archaeology* 29: 426–49.

2007. *Origins and Revolutions: Human Identity in Earliest Prehistory.* New York: Cambridge University Press.

2008. 'Kinship and Material Culture: Archaeological Implications of the Human Global Diaspora', in N. J. Allen, H. Callan, R. Dunbar and W. James, eds, *Kinship and Evolution.* Oxford: Blackwell. 27–40.

Gamble, C. S., and S. Gaudzinski. 2005. 'Bones and Powerful Individuals: Faunal Case Studies from the Arctic and European Middle Palaeolithic', in C. Gamble and M. Porr, eds, *The Individual Hominid in Context: Archaeological Investigations of Lower and Middle Palaeolithic Landscapes, Locales and Artefacts.* London: Routledge. 154–75.

Gamble, C. S., J. A. J. Gowlett and R. Dunbar. 2011. 'The Social Brain and the Shape of the Palaeolithic', *Cambridge Archaeological Journal* 21: 115–35.

Gell, A. 1998. *Art and Agency: Towards a New Anthropological Theory.* Oxford: Clarendon.
Giddens, A. 1984. *The Constitution of Society.* Berkeley: University of California Press.
Gowlett, J. A. J. 2010. 'Firing Up the Social Brain', in R. Dunbar, C. Gamble and J. A. J. Gowlett, eds, *Social Brain and Distributed Mind.* Oxford: Oxford University Press. 341–66.
Kerouac, J. 1957. *On the Road.* New York: Viking Press.
Knight, C. 1983. 'Lévi-Strauss and the Dragon: Mythologiques Reconsidered in the Light of an Australian Aboriginal Myth', *Man* 18: 21–50.
Kohn, M., and S. Mithen. 1999. 'Handaxes: Products of Sexual Selection?', *Antiquity* 73: 518–26.
Lakoff, G., and M. Johnson. 1980. *Metaphors We Live By.* Chicago: University of Chicago Press.
1999. *Philosophy in the Flesh: The Embodied Mind and Its Challenge to Western Thought.* New York: Basic.
MacLarnon, A. M., and G. P. Hewitt. 2004. 'Increased Breathing Control: Another Factor in the Evolution of Human Language', *Evolutionary Anthropology* 13: 181–97.
McPherron, S. P., Z. Alemseged, C. W. Marean, J. G. Wynn, D. Reed, D. Geraads, R. Bobe and H. A. Béarat. 2010. 'Evidence for Stone-Tool-Assisted Consumption of Animal Tissues before 3.39 Million Years Ago at Dikika, Ethiopia', *Nature* 466: 857–60.
Morris, E. L. 1994. 'The Pottery: In, Excavations at a Late Bronze Age Settlement in the Upper Thames Valley at Shorncote Quarry near Cirencester, 1992', *Transactions of the Bristol and Gloucestershire Archaeological Society* 120: 17–57.
Pettitt, P. B. 2011. *The Palaeolithic Origins of Human Burial.* London: Routledge.
Quiatt, D., and V. Reynolds. 1993. *Primate Behaviour: Information, Social Knowledge, and the Evolution of Culture.* Cambridge: Cambridge University Press.
Renfrew, C. 2007. *Prehistory: Making of the Human Mind.* London: Weidenfeld and Nicolson.
Roberts, S. G. B. 2010. 'Constraints on Social Networks', in R. Dunbar, C. Gamble and J. A. J. Gowlett, eds, *Social Brain and Distributed Mind.* Oxford: Oxford University Press. 115–34.
Rodseth, L., R. W. Wrangham, A. Harrigan and B. B. Smuts. 1991. 'The Human Community as a Primate Society', *Current Anthropology* 32: 221–54.

Said, E. W. 1978. *Orientalism: Western Conceptions of the Orient.* London: Penguin.

Semaw, S., P. Renne, J. W. K. Harris, C. S. Feibel, R. L. Bernor, N. Fesseha and K. Mowbray. 1997. '2.5 Million-Year-Old Stone Tools from Gona, Ethiopia', *Nature* 385: 333–6.

Shea, J. J. 2011. '*Homo Sapiens* Is as *Homo Sapiens* Was: Behavioural Variability versus "Behavioural Modernity" in Palaeolithic Archaeology', *Current Anthropology* 52: 1–35.

Sterner, J. 1989. 'Who Is Signalling Whom? Ceramic Style, Ethnicity and Taphonomy among the Sirak Bulahay', *Antiquity* 63: 451–9.

Taylor, J., ed. 2010. *Journey through the Afterlife: Ancient Egyptian Book of the Dead.* London: British Museum Press.

White, M. J. 1998. 'On the Significance of Acheulean Biface Variability in Southern Britain', *Proceedings of the Prehistoric Society* 64: 15–44.

Index

Index

Index

Index

Printed in the United States
By Bookmasters